Student Report Section

- <u>Title</u>: Each experiment has a title, and a space is also provided for dating the activity
- <u>Procedure</u>: The students should outline the procedures they followed in a simple list format
- <u>Diagram</u>: Diagrams should be drawn in pencil and labelled
- <u>Graphs (where necessary)</u>: Show each point clearly and enter units on each graph. Label both axes
- <u>Results</u>: Results of the experiment are entered here
- <u>Conclusions/comments</u>: Based on the results found in the activity. The students may enter their conclusion and discuss any error that may have arisen. They may also mention any social and applied aspect of the experiment
- <u>Signature</u>: By signing this the students now have their own personal record of the activity
- <u>Questions (yellow pages)</u>: These follow-up questions allow the students to test their knowledge of the experiment and its theoretical background.

Safety in the laboratory

These guidelines should be discussed in a class with a qualified person before any practical activity is undertaken. Students should remember at all times that the laboratory is a dangerous environment. Therefore, these precautions listed below should be followed:

1 Do not enter the laboratory without permission
2 Leave all bags and coats outside the working area
3 Wear safety glasses at all times
4 Always tie back long hair securely
5 Nothing must be tasted, eaten or drunk in the laboratory
6 Do not use any equipment unless permitted to do so by the teacher. If in doubt, ask the teacher
7 Always check that the label on the bottle is exactly the same as the material you require. If in doubt, ask the teacher
8 Keep all flammable chemicals away from naked flames
9 If any chemicals are splashed onto your clothing or skin, wash at once with plenty of water, and report what has happened to the teacher
10 Spit out immediately any substance accidentally taken into your mouth and wash out your mouth with plenty of water. Report what has happened to the teacher
11 Report any cut, burn or other accident immediately to the teacher
12 Always wash your hands after practical work.

Table of contents

Leaving Certificate
Chemistry
Experiment Book

Jim McCarthy ● Terence White

The Educational Company

This workbook contains a series of instructions on how to carry out the mandatory experiments prescribed by the Department of Education and Science in the Leaving Certificate Chemistry Syllabus.

The **yellow pages** are to be used during practical classes. See 'How to use this book' below.

The **white pages** are to be used after practical classes to write up a report on each of the experiments. See 'Student report section' below.

At the end of each experiment there is a selection of questions designed to test the students' knowledge of the experiment they have carried out and of a broader range of related material. An excellent knowledge of the mandatory experiments is essential for all students as Section A of the Leaving Certificate Chemistry exam (higher and ordinary) specifically examines the practical part of the course. Section A is worth from 100 to 150 marks (from 25% to 37.5% of the total marks available) depending on question choice. Questions on experimental work are usually asked in Section B also.

Practical work is an excellent and enjoyable way of investigating and strengthening theory learned in theory classes. However, it must be remembered that practical work can be potentially hazardous and students must always be aware of this.

On the next page is a list of general laboratory rules. It must be remembered that this list is not exhaustive and that any uncertainties should be discussed with the class teacher.

It is important that the students use the following guidelines while conducting, recording and assessing experimental procedures:

- Follow all steps in the manual
- Wear safety glasses for all experiments
- Outline the procedures followed
- Draw a diagram of the apparatus used
- Record the results obtained
- Draw graphs where relevant
- Evaluate results and compare to theoretical results
- Suggest experimental errors to explain differences between observed and theoretical results
- Finally, list any social and applied aspects of this experiment.

How to use this book

Each experiment follows this layout (yellow pages):

- Theory: An introduction which gives the students a background to the practical activity
- Chemicals and apparatus: This lists the chemicals and apparatus required for this experiment
- Procedure: An outline of the procedure to be followed
- Results: While carrying out the experiment the students record their initial results here
- Conclusions/comments: In this section the students make their initial conclusion on the experiment and explore any error incurred in the experiment and the reasons, if any, such errors occurred.

FLAME TESTS

Activity 1 (mandatory experiment) – Flame tests (Li, Na, K, Ba, Sr and Cu)

Method 1 - using a platinum wire

Theory:

When salts of the metals lithium, barium, potassium, copper, strontium and sodium are heated in the flame of a Bunsen burner, colours characteristic of the particular element are given off.

Chemicals and apparatus:

- Lithium chloride ☒
- Sodium chloride
- Potassium chloride
- Barium chloride ☒
- Strontium chloride
- Copper(II) chloride ☠
- Concentrated hydrochloric acid ⚗

- Platinum (or nichrome) wire held in glass rod
- Bunsen burner
- Six small beakers
- Test tubes
- Pestle and mortar
- Safety glasses

Procedure:

NB. Wear your safety glasses.

1   Clean the platinum wire using concentrated hydrochloric acid in a test tube. Do this in the fume cupboard.

2   Crush the salt to be tested with a pestle and mortar, and transfer it to a labelled beaker.

3   Dip the platinum wire in concentrated hydrochloric acid and then in the salt to be tested.

4   Place the platinum wire in the blue flame of the Bunsen burner as in the diagram, and note the colour given off.

5   Repeat the experiment for each of the other salts. Again, note the colour in each case.

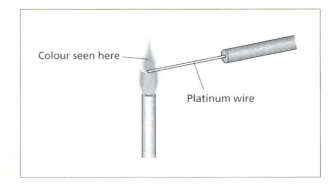

Method 2 – using a soaked wooden splint

Chemicals and apparatus:

- Lithium chloride ✖
- Sodium chloride
- Potassium chloride
- Barium chloride ✖
- Strontium chloride
- Copper(II) chloride ☠

- Soaked wooden splints
- Bunsen burner
- Six small beakers
- Test tubes
- Pestle and mortar
- Safety glasses

Procedure:

NB. Wear your safety glasses.

1 You will need a soaked wooden splint for each sample to avoid cross-contamination.

2 Crush the salt to be tested with a pestle and mortar, and transfer it to a labelled beaker.

3 Dip the soaked splint in the salt to be tested.

4 Place the splint in the blue flame of the Bunsen burner and note the colour given off.

5 Repeat the experiment for each of the other salts. Again, note the colour in each case.

Metal	Salt being tested	Flame colour
Barium		
Copper		
Lithium		
Potassium		
Sodium		
Strontium		

Conclusions/comments:

_____

_____

_____

Title:_____**Date:**_____

Chemicals and apparatus:

_____

_____

_____

Procedure:

_____

_____

_____

_____

_____

_____

_____

_____

_____

_____

Results:

Metal	Salt tested	Flame colour
Barium		
Copper		
Lithium		
Potassium		
Sodium		
Strontium		

Conclusions/comments:

_____

_____

_____

_____

Signature:_____

I   Why is concentrated hydrochloric acid used in this experiment (Method 1)?

_____

2   Cross contamination could be a problem in this experiment. What can you do to avoid cross contamination?

_____

_____

_____

3   When a sample of an unknown metal salt was heated in a flame, a blue-green colour was observed. What metal was present in the salt?

_____

4   Why do metal salts give off a characteristic colour in a flame?

_____

_____

5   Complete the following: When a sample of a sodium salt is heated on a platinum wire in a blue flame, a _____ colour characteristic of the metal is observed, while if a sample of a potassium salt is used instead, a _____ colour characteristic of the metal is observed.

6   You are given a sample of an unknown metal salt. Explain how you would find out what metal is present in the salt.

_____

_____

_____

7   Complete the table below:

Metal	Salt tested	Flame colour
	Sodium chloride	
		Deep red
		Yellow-green
Potassium		
Copper		
	Strontium chloride	

8   State one safety precaution that you would take for this experiment.

_____

OXIDATION AND REDUCTION EXPERIMENTS

Activity 2a (mandatory experiment) – Redox reactions of the halogens

Theory:

The halogens chlorine, bromine and iodine are very reactive elements. They often react by taking an electron from another element or ion. This means that they act as **oxidising agents**. The smaller the halogen atom, the stronger the oxidising agent it is. In terms of oxidising power:

Cl > Br > I.

(i) Reactions with halides

- Chlorine solution ☠ ✖ i
- Bromine solution ☠ 🜢
- Iodine solution
- Sodium chloride solution
- Sodium bromide solution
- Potassium iodide solution
- Acidified iron(II) sulfate solution ✖ n
- Iron(III) chloride solution ✖ i
- Sodium sulfite solution
- Sodium hydroxide solution 🜢
- Silver nitrate solution

- Barium chloride solution ✖ n
- Dilute hydrochloric acid 🜢
- Dilute ammonia solution ✖ i
- PVC gloves
- Fume cupboard or well-ventilated room
- Pasteur pipettes
- Test tubes
- Test tube rack
- Test tube brush
- Safety glasses

Chemicals and apparatus:

Theory:

Chlorine is a stronger oxidising agent than bromine and iodine, and is capable of releasing these elements from solutions of their salts:

$$Cl_{2(aq)} + 2Br^-_{(aq)} \rightarrow 2Cl^-_{(aq)} + Br_{2(aq)}$$
$$Cl_{2(aq)} + 2I^-_{(aq)} \rightarrow 2Cl^-_{(aq)} + I_{2(aq)}$$

Bromine can release iodine from a solution of its salts:

$$Br_{2(aq)} + 2I^-_{(aq)} \rightarrow 2Br^-_{(aq)} + I_{2(aq)}$$

Procedure:

NB. Wear your safety glasses.

I   Look at each of the following solutions: chlorine solution, bromine solution, iodine solution, sodium chloride solution, sodium bromide solution, potassium iodide solution. Record the colour of each solution in Table 1.

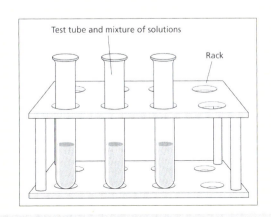

Test tube and mixture of solutions

Rack

2 Add 2 cm³ of the chlorine solution and the sodium bromide solution respectively to separate test tubes and mix. Record your observations in Table 2.

3 Add 2 cm³ of the chlorine solution and the potassium iodide solution respectively to separate test tubes and mix. Record your observations in Table 2.

4 Add 2 cm³ of the bromine solution and the potassium iodide solution respectively to separate test tubes and mix. Record your observations in Table 2.

Solution	Colour of solution
Chlorine in water	
Bromine in water	
Iodine in water	
Chloride ions in water	
Bromide ions in water	
Iodide ions in water	

Solutions added to the test-tube	Observation
(a) Chlorine and bromide ions	
(b) Chlorine and iodide ions	
(c) Bromine and iodide ions	

Conclusions/comments:

_____

_____

_____

_____

(ii) Reactions with iron(II) salts and with sulfites

Theory:

Solutions of chlorine, bromine and iodine are all able to oxidise iron(II) ions to iron(III) ions, and to oxidise sulfite ions to sulfate ions in aqueous solution.

Chlorine reacts with iron(II) ions as follows:

$$Cl_{2(aq)} + 2Fe^{2+}_{(aq)} \rightarrow 2Cl^-_{(aq)} + 2Fe^{3+}_{(aq)}$$

Chlorine reacts with sulfite ions as shown:

$$Cl_{2(aq)} + SO_3^{2-}_{(aq)} + H_2O_{(l)} \rightarrow 2Cl^-_{(aq)} + SO_4^{2-}_{(aq)} + 2H^+_{(aq)}$$

Bromine reacts with iron(II) ions as follows:

$$Br_{2(aq)} + 2Fe^{2+}_{(aq)} \rightarrow 2Br^-_{(aq)} + 2Fe^{3+}_{(aq)}$$

Bromine reacts with sulfite ions as shown:

$$Br_{2(aq)} + SO_3^{2-}_{(aq)} + H_2O_{(l)} \rightarrow 2Br^-_{(aq)} + SO_4^{2-}_{(aq)} + 2H^+_{(aq)}$$

Iodine reacts with iron(II) ions as follows:

$$I_{2(aq)} + 2Fe^{2+}_{(aq)} \rightarrow 2I^-_{(aq)} + 2Fe^{3+}_{(aq)}$$

Iodine reacts with sulfite ions as shown:

$$I_{2(aq)} + SO_3^{2-}_{(aq)} + H_2O_{(l)} \rightarrow 2I^-_{(aq)} + SO_4^{2-}_{(aq)} + 2H^+_{(aq)}$$

To find out whether sulfate ions are formed, barium chloride solution and hydrochloric acid are used. Both sulfite and sulfate ions react with barium chloride solution, producing white precipitates of barium sulfite and barium sulfate respectively. Barium sulfite dissolves in hydrochloric acid but barium sulfate does not. This reaction is used to distinguish between sulfite and sulfate ions.

Procedure:

NB. Wear your safety glasses.

1. Look at each of the following solutions: chlorine solution, iron(II) sulfate solution, iron(III) chloride solution, iron(II) sulfate solution in sodium hydroxide solution, iron(III) chloride solution in sodium hydroxide solution. Record the colour of each solution in Table 3.

2. Add 2 cm³ of the chlorine solution and the iron(II) sulfate solution respectively to separate test tubes and mix. Then add 10 drops of the sodium hydroxide solution to the mixture. Record your observations in Table 4.

3. Add 2 cm³ of the chlorine solution and the sodium sulfite solution respectively to separate test tubes and mix. Using a dropping pipette add a few drops of barium chloride solution. Now add 2 cm³ of dilute hydrochloric acid. Record your observations in Table 4.

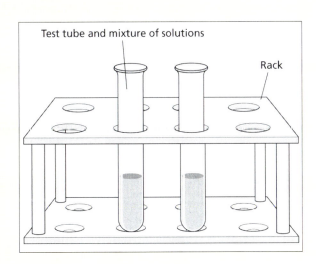

Results:

Table 3

Solution	Colour of solution
Chlorine in water	
Iron(II) sulfate in water	
Iron(III) chloride in water	
Iron(II) sulfate solution in sodium hydroxide solution	
Iron(III) chloride solution in sodium hydroxide solution	

Table 4

Solutions added to the test-tube	Observation
Chlorine and iron(II) sulfate solutions followed by 10 drops of sodium hydroxide solution	
Chlorine and sodium sulfite solutions followed by the test for the presence of sulfate ions	

Conclusions/comments:

_____

_____

_____

_____

Student report, Activity 2a

Title:_____**Date:**_____

Chemicals and apparatus:

_____

_____

_____

Procedure:

_____

_____

_____

_____

_____

_____

_____

_____

_____

_____

_____

_____

_____

_____

_____

_____

_____

_____

_____

_____

_____

_____

_____

_____

_____

_____

Results:

Solutions added to the test tube	Observation	Conclusion
(a) Chlorine and bromide ions		
(b) Chlorine and iodide ions		
(c) Bromine and iodide ions		
(d) Chlorine and iron(II) sulfate solutions followed by 10 drops of sodium hydroxide solution		
(e) Chlorine and sodium sulfite solutions followed by the test for the presence of sulfate ions		

Conclusions/comments:

_____

_____

_____

_____

Signature:_____

Student report, Activity 2a – diagram.

Activity 2b (mandatory experiment) – Displacement reactions of metals

Theory:

A metal will displace a less reactive metal, that is, one which is below it in the electrochemical series, from a solution of its salts. The more reactive metal is oxidised and forms a water-soluble positive ion. The less reactive metal is reduced and the solid metal is formed.

In this experiment, zinc and magnesium respectively are reacted with a solution of copper(II) sulfate:

$$Zn_{(s)} + CuSO_{4(aq)} \rightarrow ZnSO_{4(aq)} + Cu_{(s)}$$

$$Mg_{(s)} + CuSO_{4(aq)} \rightarrow MgSO_{4(aq)} + Cu_{(s)}$$

This experiment works best under acidic conditions; under these conditions, the following reactions take place simultaneously:

$$Zn_{(s)} + H_2SO_{4(aq)} \rightarrow ZnSO_{4(aq)} + H_{2(g)}$$

$$Mg_{(s)} + H_2SO_{4(aq)} \rightarrow MgSO_{4(aq)} + H_{2(g)}$$

Chemicals and apparatus:

- Acidified copper(II) sulfate solution
- Zinc powder
- Magnesium ribbon
- Pasteur pipettes
- Boiling tubes
- Boiling tube rack
- Test tube brush
- Safety glasses

Procedure:

NB. Wear your safety glasses.

Results:

1   Half fill two boiling tubes with the acidified copper(II) sulfate solution.

2   Add the magnesium ribbon to the solution in one boiling tube. Record your observations.

3   Add the zinc powder to the solution in the other boiling tube. Record your observations.

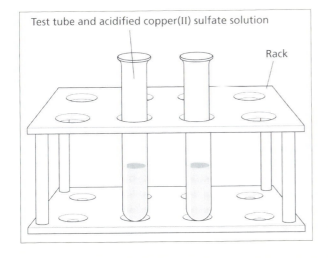

Test tube and acidified copper(II) sulfate solution

Rack

Results:

Metal	Magnesium	Zinc
(a) Colour of copper(II) sulfate solution at the beginning		
(b) Colour of the solution at the end of the reaction		
(c) Colour of the precipitate formed		
(d) Any other observations		

Conclusions/comments:

_____

_____

_____

_____

Title:_____ **Date:**_____

Chemicals and apparatus:

_____

_____

_____

_____

Procedure:

_____

_____

_____

_____

_____

_____

_____

Results:

Metal	Magnesium	Zinc
(a) Colour of copper(II) sulfate solution at the beginning		
(b) Colour of the solution at the end of the reaction		
(c) Colour of the precipitate formed		
(d) Any other observations		

Metal	Observation	Conclusion
Mg	(a)	
	(b)	
	(c)	
	(d)	
Cu	(a)	
	(b)	
	(c)	
	(d)	

Conclusions/comments:

_____

_____

_____

Signature:_____

Student report, Activity 2b – diagram

1   When chlorine solution is added to potassium iodide solution, the reddish-brown colour of iodine appears. Write an equation for this reaction, and state what is oxidised and what is reduced.

_____

_____

_____

2   What is observed when chlorine solution is added to a solution containing bromide ions? Write an equation for the reaction.

_____

_____

_____

3   Chlorine solution reacts with sodium bromide solution, but if bromine is mixed with sodium chloride solution, no reaction occurs. Explain.

_____

_____

_____

4   Why is the iron(II) sulfate solution used in this experiment acidified?

_____

_____

5   Describe the test used to show that Fe^{3+} ions are formed in a reaction in this experiment.

_____

_____

6   What colour change would you expect to happen if sodium sulfite solution is added to iodine solution? Explain your answer.

_____

_____

_____

7   What two products are formed when acidified iron(II) sulfate solution is added to a solution of bromine?

_____

_____

8   If zinc is added to a solution of acidified copper(II) sulfate, the solution becomes lighter in colour, and the zinc becomes covered with copper. Explain.

_____

_____

9 When magnesium is added to copper(II) sulfate solution, the copper is displaced according to the equation:

Mg + CuSO$_4$ → Cu + MgSO$_4$

(i) State **one** change observed as the reaction proceeds.

_____

(ii) Which substance is oxidised?

_____

10 What would you expect to happen if magnesium is added to a solution of zinc chloride? Explain your answer.

_____

_____

_____

CHAPTER 3
ANION TESTS

Activity 3 (mandatory experiment) – Tests for anions in aqueous solutions: chloride Cl^-, carbonate CO_3^{2-}, nitrate NO_3^-, sulfate SO_4^{2-}, phosphate PO_4^{3-}, sulfite SO_3^{2-}, hydrogencarbonate HCO_3^-

Theory:

Negatively charged ions, anions, may be identified by tests carried out on their aqueous solutions. Easily identifiable results, which may include the production of characteristically coloured precipitates, are obtained when the aqueous solutions react with certain reagents.

Chemicals and apparatus:

- Deionised water
- Dilute hydrochloric acid
- Limewater ✖ i
- Magnesium sulfate solution
- Barium chloride solution ✖ n
- Silver nitrate solution
- Dilute ammonia solution
- Cold saturated solution of iron(II) sulphate
- Concentrated sulfuric acid
- Disodium hydrogen phosphate solution
- Ammonium molybdate reagent ✖ n
- Dilute solutions of the salts being tested

- Test tubes
- Test tube rack
- Test tube holder
- Stoppers for test tubes fitted with plastic delivery tubing
- Labels
- Bunsen burner
- Droppers
- Beakers
- Wash bottle
- Thermometer
- Safety glasses

Activity 3a (mandatory experiment) – To test for the carbonate (CO_3^{2-}) and hydrogencarbonate (HCO_3^-) anions

Theory:

Carbonate and hydrogencarbonate ions both react with dilute hydrochloric acid producing carbon dioxide gas:

(i) $Na_2CO_{3(aq)} + 2HCl_{(aq)} \rightarrow 2NaCl_{(aq)} + H_2O_{(l)} + CO_{2(g)}$

(ii) $NaHCO_{3(aq)} + HCl_{(aq)} \rightarrow NaCl_{(aq)} + H_2O_{(l)} + CO_{2(g)}$

To distinguish between carbonate and hydrogencarbonate ions, a solution of magnesium sulfate is added. Since magnesium carbonate is insoluble in water, a white precipitate indicates that the salt is a carbonate:

$Na_2CO_{3(aq)} + MgSO_{4(aq)} \rightarrow Na_2SO_{4(aq)} + MgCO_{3(s)}$

The absence of a precipitate indicates a hydrogencarbonate because magnesium hydrogencarbonate is soluble in water:

$2NaHCO_{3(aq)} + MgSO_{4(aq)} \rightarrow Na_2SO_{4(aq)} + Mg(HCO_3)_{2(aq)}$

Procedure:

NB. Wear your safety glasses.

1 Add 2 cm³ of a carbonate solution to a test tube.

2 Add some dilute hydrochloric acid using a dropper. Record what happens.

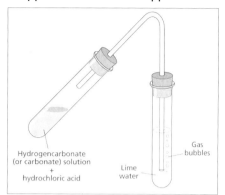

3 Repeat steps 1 and 2 using the arrangement shown in the diagram above. Record what happens.

4 Add a few cm³ of magnesium sulfate solution to some fresh carbonate solution in a clean test tube. Record what happens.

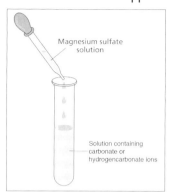

5 Repeat steps 1 to 4, using a hydrogencarbonate solution instead of a carbonate solution.

Results:

	Carbonate	Hydrogencarbonate
Observations on adding dilute hydrochloric acid		
Observations and conclusion from limewater test		
Observation on adding magnesium sulfate solution		

Conclusions/comments:

_____

_____

_____

_____

Title:_____**Date:**_____

Chemicals and apparatus:

_____

_____

_____

Procedure:

_____

_____

_____

_____

_____

_____

_____

_____

_____

_____

_____

Results:

	Carbonate	Hydrogencarbonate
Observations on adding dilute hydrochloric acid		
Observations and conclusion from limewater test		
Observation on adding magnesium sulfate solution		

Conclusions/comments:

_____

_____

_____

_____

Signature:_____

Activity 3b (mandatory experiment) – To test for the sulfite (SO_3^{2-}) and sulfate (SO_4^{2-}) anions

Theory:

Both sulfite and sulfate ions react with barium chloride solution, producing white precipitates of barium sulfite and barium sulfate respectively:

(i) $Na_2SO_{3(aq)} + BaCl_{2(aq)} \rightarrow 2NaCl_{(aq)} + BaSO_{3(s)}$

(ii) $Na_2SO_{4(aq)} + BaCl_{2(aq)} \rightarrow 2NaCl_{(aq)} + BaSO_{4(s)}$

Barium sulfite reacts and dissolves in hydrochloric acid but barium sulfate does not:

$$BaSO_{3(s)} + 2HCl_{(aq)} \rightarrow BaCl_{2(aq)} + H_2O_{(l)} + SO_{2(g)}$$

This reaction is used to distinguish between sulfite and sulfate ions.

Procedure:

NB. Wear your safety glasses.

1   Add 2 cm³ of a sulfite solution to a clean test tube and add some barium chloride solution using a dropper. Record what happens.

2   Add a few cm³ of hydrochloric acid and mix gently. Record what happens.

3   Repeat steps 1 and 2, using a sulfate solution instead of a sulfite solution.

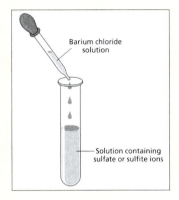

Barium chloride solution

Solution containing sulfate or sulfite ions

Results:

	Sulfite	**Sulfate**
Observation on adding barium chloride solution		
Observation on adding hydrochloric acid		

Conclusions/comments:

_____

_____

_____

_____

Title:_____**Date:**_____

Chemicals and apparatus:

_____

_____

_____

Procedure:

_____

_____

_____

_____

_____

_____

_____

Results:

	Sulfate	Sulfite
Observation on adding barium chloride solution		
Observation on adding hydrochloric acid		

Conclusions/comments:

_____

_____

_____

_____

Signature:_____

Activity 3c (mandatory experiment) – To test for the chloride (Cl⁻) anion

Theory:

Solutions containing chloride ions react with silver nitrate solution, producing a white precipitate of silver chloride:

This precipitate dissolves on addition of dilute ammonia solution.

$$NaCl_{(aq)} + AgNO_{3(aq)} \rightarrow NaNO_{3(aq)} + AgCl_{(s)}$$

Procedure:

NB. Wear your safety glasses.

1　Add 2 cm³ of chloride solution to a clean test tube.

2　Add a few drops of silver nitrate solution using a dropper. Record what happens.

3　Add a few cm³ of dilute ammonia solution. Record what happens.

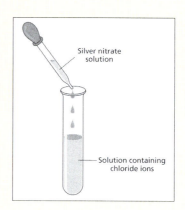

Results:

	Chloride
Observations on adding silver nitrate solution	
Observation on adding dilute ammonia solution	

Conclusions/comments:

_____

_____

_____

_____

Student report, Activity 3c

Title: ______________________ **Date:** __________

Chemicals and apparatus:

Procedure:

Results:

	Chloride
Observations on adding silver nitrate solution	
Observation on adding dilute ammonia solution	

Conclusions/comments:

Signature: ______________________

Activity 3d (mandatory experiment) – To test for the nitrate NO₃⁻ anion

Theory:

Solutions containing nitrate ions react with a mixture of iron(II) sulfate solution and concentrated sulfuric acid. When concentrated sulfuric acid is added to a mixture of nitrate and iron(II) sulfate solutions, a brown ring develops slowly at the interface of the sulfuric acid layer and the layer containing the mixture.

Procedure:

NB. Wear your safety glasses.

1   Add 2 cm³ of potassium nitrate solution to a clean test tube.

2   Add 3 cm³ of cold saturated iron(II) sulfate solution using a dropper. Record what happens.

3   Carefully add 2 cm³ of concentrated sulfuric acid slowly down the wall of the test tube using a dropper. Do not mix the contents of the test tube.

4   Record what happens after a few minutes.

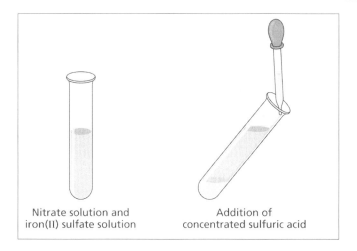

Nitrate solution and iron(II) sulfate solution

Addition of concentrated sulfuric acid

Results:

	Nitrate
Observation on adding iron(II) sulfate solution	
Observation on adding concentrated sulfuric acid	

Conclusions/comments:

_____

_____

_____

_____

Student report, Activity 3d

Title:____ **Date:**____________________

Chemicals and apparatus:

Procedure:

Results:

	Nitrate
Observation on adding iron(II) sulfate solution	
Observation on adding concentrated sulfuric acid	

Conclusions/comments:

Signature: ______________________________

Activity 3e (mandatory experiment) – To test for the phosphate (PO_4^{3-}) anion

Theory:

Solutions containing phosphate ions react on heating with an ammonium molybdate reagent, forming a yellow precipitate. This precipitate dissolves on addition of ammonia solution.

Procedure:

NB. Wear your safety glasses.

1. Add 2 cm³ of disodium hydrogenphosphate(V) solution to a clean test tube.

2. Add about 6 cm³ of the clear ammonium molybdate reagent to the test tube using a dropper.

3. Warm gently by placing in a beaker of water at a temperature not exceeding 40°C. Record what happens.

4. To the contents of the test tube at the end of step 3, add an equal volume of ammonia solution. Record what happens.

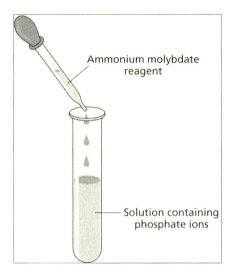

Ammonium molybdate reagent

Solution containing phosphate ions

Results:

	Phosphate
Observations on adding ammonium molybdate reagent and heating	
Observations on adding ammonia solution	

Conclusions/comments:

_____

_____

_____

_____

Title:_____**Date:**_____

Chemicals and apparatus:

_____

_____

_____

Procedure:

_____

_____

_____

_____

_____

_____

_____

_____

Results:

	Phosphate
Observations on adding ammonium molybdate reagent and heating	
Observations on adding ammonia solution	

Conclusions/comments:

_____

_____

_____

_____

Signature:_____

1 Why is deionised water, and not tap water, used in all of these experiments?

_____

_____

2 What is observed when a solution of silver nitrate is added to a solution of sodium chloride? Why does this happen?

_____

_____

_____

3 What gas is given off when hydrochloric acid is added to a solution of hydrogencarbonate ions? What solution can be used to confirm this?

_____

_____

_____

4 If magnesium sulfate solution is added to a solution containing hydrogencarbonate ions, nothing happens, but if the mixture is then heated, a white precipitate is formed. Explain.

_____

_____

_____

5 You are given samples of the following salts: sodium sulfate (Na_2SO_4) and sodium sulfite (Na_2SO_3). Describe the test which could be carried out to distinguish between the sulfate salt and the sulfite salt.

_____

_____

_____

_____

6 In testing for the presence of nitrate ions in aqueous solution, particular care must be taken. Why is this?

_____

_____

_____

7 Complete the following table:

Test for	Reagents needed	Positive result
Chloride	Silver nitrate solution Ammonia solution	
		Brown ring
Sulfate	Barium chloride solution Hydrochloric acid	
Phosphate		
Carbonate	Hydrochloric acid Magnesium sulfate solution	
Hydrogencarbonate		

Activity 4 (mandatory experiment) – Recrystallisation of benzoic acid and determination of its melting point

(a) Recrystallisation

Theory:

An impure solid may be purified by first dissolving it in the minimum quantity of a boiling solvent. The hot mixture is then filtered rapidly to remove insoluble impurities. Crystals will form as the filtrate is allowed to cool slowly, since the solubility of the solid decreases as the temperature falls.

Chemicals and apparatus:

- Benzoic acid (impure sample)
- Deionised water
- Ice-cold deionised water
- Beakers (100 cm^3)
- Beaker (250 cm^3)
- Conical flask (250 cm^3)
- Filter funnel
- Filter paper
- Clock glass
- Buchner funnel and filter papers to fit

- Buchner flask and rubber adaptor
- Retort stand and clamp
- Filter pump (or vacuum pump)
- Spatula
- Hotplate/magnetic stirrer or Bunsen burner, tripod and gauze
- Magnetic pellet
- Tongs
- Heat resistant mat
- Safety glasses

Procedure:

NB. Wear your safety glasses.

1. Boil about 100 cm^3 of deionised water in a 250 cm^3 beaker.
2. Keep the water just at its boiling point.
3. Flute a filter paper, place it in the funnel, and wet it.
4. Place the funnel (with fluted filter paper) in the neck of a 250 cm^3 conical flask containing about 20 cm^3 of boiling water, and standing on a hotplate.
5. Place about 1 g of the impure benzoic acid sample in a small beaker.
6. Dissolve the benzoic acid in the minimum amount of boiling water.

7. Stand the beaker on a hot plate to keep the solution near its boiling point.

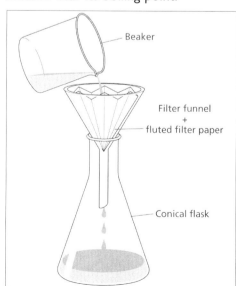

8 Empty the conical flask, replace the funnel quickly, and stand the flask on a heat resistant mat.

9 Pour the boiling solution through the filter paper in small portions.

10 If any crystals form in the filter paper, add a little boiling water to dissolve them.

11 Pour the filtrate into a warm 100 cm³ beaker.

12 Evaporate off the water until traces of crystals begin to appear on the sides of the beaker.

13 Allow the filtrate to cool to room temperature.

14 Carefully filter the recrystallised mixture, by vacuum filtration if possible.

15 Wash the crystals with small portions of ice-cold deionised water.

16 Allow the crystals to dry.

Conclusions/comments:

_____

_____

_____

_____

(b) Melting point determination

Theory:

The melting point range of a substance is the narrow band of temperatures between the temperature at which melting begins and the temperature at which the entire solid has liquefied.

The melting point range of an impure substance is lower and wider than that of a pure sample of the same substance. Comparison of the melting point ranges of the purified sample with pure benzoic acid gives an indication of the success of the purification process.

Chemicals and apparatus:

- Impure sample of benzoic acid
- Purified sample of benzoic acid
- Sample of pure benzoic acid
- Melting point tubes
- Pestle and mortar
- Bunsen burner or hotplate
- Aluminium block with hole drilled for thermometer or alternative melting point apparatus

- Boiling tubes
- Retort stand and clamp
- Thermometer (preferably -10 → 150 °C)
- Heat-resistant gloves
- Safety glasses

Method 1 – using an aluminium block

Procedure:

NB. Wear your safety glasses.

1   Place a few crystals of the substance to be melted on the surface of a clean aluminium block into which a thermometer has been placed.

2   Use a Bunsen burner or hot plate to heat the block slowly.

3   Record the melting point range i.e. the temperature at which melting begins and the temperature at which it ends.

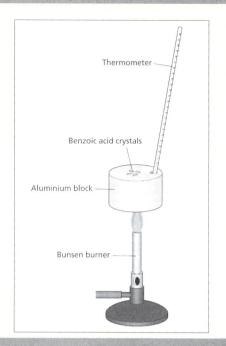

Method 2 – using an oil bath

Procedure:

NB. Wear your safety glasses.

1   Grind the sample in a mortar.

2   Fill a melting point tube to a depth of 3–5 mm with the crystals.

3   Using a rubber band, attach the melting point tube to the thermometer so that the sample is near the bulb.

4   Clamp it so that the sample is well below the surface of liquid paraffin contained in the boiling tube.

5   Heat the boiling tube gently with a Bunsen burner.

6   Record the melting point range i.e. the temperature at which melting begins and the temperature at which it ends.

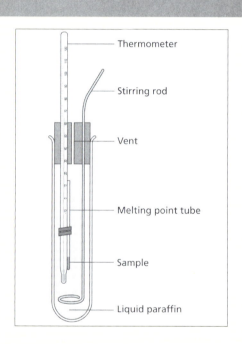

Results:

Sample	Melting point value (specify units)
Impure benzoic acid	
Purified benzoic acid	
Pure benzoic acid	

Conclusions/comments:

_____

_____

_____

_____

(a) Recrystallisation

Title:_____**Date:**_____

Chemicals and apparatus:

_____

_____

_____

Procedure:

_____

_____

_____

_____

_____

_____

_____

_____

_____

_____

_____

_____

_____

_____

_____

_____

_____

_____

_____

_____

_____

_____

_____

_____

_____

_____

Conclusions/comments:

_____

_____

_____

_____

Signature:_____

Student report, Activity 4 – diagram

(a) Recrystallisation

(b) Melting point determination

Title:_____ **Date:**_____

Chemicals and apparatus:

_____

_____

_____

Procedure:

_____

_____

_____

_____

_____

_____

_____

_____

_____

_____

_____

_____

Results:

Sample	Melting point value (specify units)
Impure benzoic acid	
Purified benzoic acid	
Pure benzoic acid	

Conclusions/comments:

_____

_____

_____

_____

Signature:_____

(b) Melting point determination

1 Why might water be an unsuitable solvent for the recrystallisation of certain compounds?

_____

_____

2 What would be the effect on the yield of crystals of benzoic acid if more than the minimum amount of solvent were used?

_____

3 What action could be taken before recrystallisation if too much solvent had been used?

_____

4 What action could be taken if some of the crystals formed in the leg of the funnel or on the filter paper during the hot filtration?

_____

5 How could you ensure that recrystallisation was complete?

_____

6 How could the crystals be dried?

_____

7 When measuring melting point, why is it important to heat the apparatus slowly?

_____

8 In your experiment, did the pure or impure samples of benzoic acid have the lower melting point? Explain this result.

_____

_____

9 Do the results of your experiment indicate that benzoic acid is an ionic compound or a covalent compound? Explain your answer.

_____

_____

10 If pure samples of two different substances with the same melting points were mixed, and the melting point of the mixture recorded, would you expect the result to be higher, lower or the same as that of the original substances? Explain your reasoning.

_____

_____

11 Two samples have melting points 112–116 °C and 120–121 °C. Could they be the same substance? How would you check your answer?

_____

_____

DETERMINATION OF RELATIVE MOLECULAR MASS

Activity 5 (mandatory experiment) – Estimation of the relative molecular mass, M_r, of a volatile liquid

Theory:

The relative molecular mass of a volatile liquid can be determined by two different methods, involving a conical flask and a gas syringe respectively. A small amount of the liquid is heated so that its vapour fills the flask or the gas syringe. The pressure, volume, mass and temperature of the vapour are measured, and, using the equation of state of an ideal gas, **PV= nRT**, the number of moles present and the relative molecular mass can be calculated.

Method 1 – using a conical flask

Chemicals and apparatus:

- Propanone 🔥
- Water
- Conical flask (250 cm³)
- Beaker (600 cm³ – into which the 250 cm³ conical flask can be easily fitted)
- Aluminium foil
- Clamp
- Dropping pipette

- Pin
- Rubber band
- Bunsen burner, tripod and gauze
- Thermometer
- Barometer
- Electronic balance
- Graduated cylinders (100 cm³)
- Safety glasses

Procedure:

NB. Wear your safety glasses.

1. Two-thirds fill the beaker with water, place on the tripod and heat to almost boiling with the Bunsen burner.

2. Control the flame so that the temperature remains at about 95 °C.

3. Cut a circle of aluminium foil a little more than large enough to cover the mouth of the clean, dry, conical flask and fold down around the sides of the flask.

4. Find the total mass of the conical flask, the aluminium foil and the rubber band.

5. Using a dropping pipette, add 3 to 4 cm³ of propanone to the flask.

6. Cover the mouth of the flask with the aluminium foil.

7. Secure the foil tightly with the rubber band so that no vapour can escape.

8. With the pin, prick one small hole in the centre of the aluminium foil cap.

9. Attach the clamp to the neck of the flask.

10. Immerse the conical flask in the hot water.

11. Holding the clamp, move the flask up and down periodically to check the liquid level in the flask.

12. The propanone will vaporise and some of it will escape out through the hole in the cap. When the flask appears to be empty (i.e. all the liquid appears to have evaporated), immediately remove the flask from the beaker.

13. Record the exact temperature of the hot water.

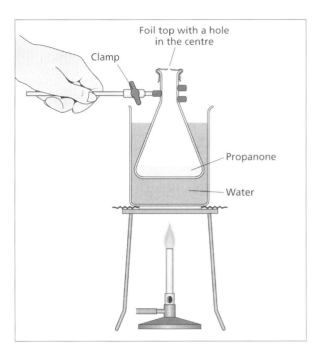

14 Allow the flask to cool – most of the vapour will condense.

15 Thoroughly dry the outside of the flask, including the foil.

16 Find the mass of the flask, cap, rubber band and contents.

17 Remove the cap and rubber band.

18 Find the volume of the flask by completely filling it with water, and then transferring all of the liquid from it to graduated cylinders.

19 Record the volume of the liquid transferred.

20 Observe the value of atmospheric pressure using the barometer, and record this value.

Results:

(Note that 760 mmHg = 101325 Pa)

	Values (specify units)
Mass of flask, cap and rubber band	
Mass of condensed vapour, flask, cap and rubber band	
Mass of condensed vapour	
Temperature of boiling water	
Temperature of boiling water in kelvins	
Volume of flask	
Volume of flask in m^3	
Atmospheric pressure	
Atmospheric pressure in Pa	
Gas constant (R)	8.31 J K^{-1} mol^{-1}

Calculations:

P $\quad = \quad$ Pa

V $\quad = \quad$ m^3

R $\quad = \quad$ 8.31 J K^{-1} mol^{-1}

T $\quad = \quad$ K

n = PV / RT $\quad = \quad$ moles

m $\quad = \quad$ g

M_r = m / n $\quad = \quad$

Conclusions/comments:

_____

_____

_____

_____

Method 2 – using a gas syringe

Chemicals and apparatus:

- Propanone 🔥
- Water
- Heat resistant gas syringe (100 cm³)
- Self-sealing rubber cap for gas syringe
- Hypodermic syringe (5 cm³) and needle

- Gas syringe heater (electrical)
- Thermometer
- Barometer
- Electronic balance
- Safety glasses

Procedure:

NB. Wear your safety glasses.

1   Put the gas syringe and the thermometer into the heater.

2   Draw about 5 cm³ of air into the gas syringe and seal with the rubber cap.

3   Switch on the heater, set it to 100°C and allow it to heat up.

4   Allow time for the temperature to equilibrate inside the heater.

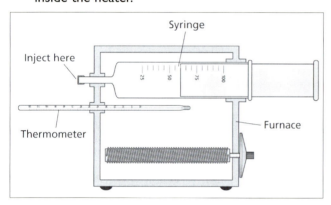

5   Record the volume of hot air in the gas syringe.

6   Draw about 0.2 cm³ of propanone into the hypodermic syringe via the needle.

7   Find the mass of the syringe, needle and contents.

8   Inject the contents of the hypodermic syringe into the gas syringe through the rubber cap.

9   Withdraw the needle and the cap reseals.

10   Ensure that the narrow neck (stem) of the syringe is inside the oven, and that only the rubber cap protrudes. The propanone will vaporise and expand inside the gas syringe, pushing on the plunger of the gas syringe. When the pressure inside the syringe is equal to the atmospheric pressure outside, the plunger will come to rest.

11   Find the mass of the hypodermic syringe and needle after the injection.

12   Record the total volume of hot air and vapour in the gas syringe.

13   Record the temperature inside the heater.

14   Observe the value of atmospheric pressure using the barometer, and record this value.

Results:

(Note that 760 mmHg = 101325 Pa)

	Values (specify units)
Mass of syringe, needle and contents before injection	
Mass of syringe, needle and contents after injection	
Mass of vapour	
Volume of heated air	
Volume of heated air and vapour	
Volume of vapour	
Volume of vapour in m³	
Temperature inside heater	
Temperature of vapour in kelvins	
Atmospheric pressure	
Atmospheric pressure in Pa	
Gas constant (R)	8.31 J K⁻¹ mol⁻¹

Calculations:

P = Pa

V = m³

R = 8.31 J K⁻¹ mol⁻¹

T = K

$n = PV / RT$ = moles

m = g

$M_r = m / n$ =

Conclusions/comments:

_____

_____

_____

_____

Title:_____**Date**:_____

Chemicals and apparatus:

_____

_____

_____

Procedure:

_____

_____

_____

_____

_____

_____

_____

_____

_____

_____

_____

_____

_____

_____

_____

_____

_____

_____

_____

_____

_____

_____

_____

_____

_____

_____

_____

_____

_____

_____

Results:

	Values (specify units)

Calculations

P	=	Pa
V	=	m^3
R	=	8.31 J K^{-1} mol^{-1}
T	=	K
n = PV / RT	=	moles
m	=	g
M_r = m / n	=	

Conclusions/comments:

_____

_____

_____

_____

Signature:_____

1   Why is it important for the conical flask to be dry before adding the volatile liquid (Method 1)?

_____

_____

2   Why is it necessary to make a small hole in the aluminium cap (Method 1)?

_____

_____

3   Why must the outside of the flask be dried completely before weighing it at the end of the experiment (Method 1)?

_____

_____

4   If a small drop of water were present in the gas syringe used in Method 2, how would this affect the results? Explain your answer.

_____

_____

5   Why is this experiment only suitable for measuring M_r of **volatile** liquids?

_____

_____

6   Give two examples of liquids that would be suitable for use in this experiment.

_____

7   Glycerol is an example of a liquid which is not volatile. Name a technique that could be used to measure its relative molecular mass.

_____

8   What is an ideal gas?

_____

9   Would a real gas behave more like an ideal gas at high temperature or at low temperature?

_____

10   Would a real gas behave more like an ideal gas at high pressure or at low pressure?

_____

11   In the equation of state for ideal gases PV = nRT, what is represented by (i) n, (ii) R?

(i) _____ (ii)_____

Activity 6 (mandatory experiment) – Preparation of a standard solution of sodium carbonate

Theory:

A standard solution is a solution whose concentration is accurately known. Anhydrous sodium carbonate (Na_2CO_3) is a highly pure and stable substance that can be used to make up a standard solution. It has a molar mass of 106 g mol^{-1}. A 0.1 M solution is made up, using a 250 cm^3 volumetric flask. For 250 cm^3 of 0.1 M sodium carbonate solution, the mass required is:

$$106 \times 0.1 \times 250 / 1000 = 2.65 \text{ g}$$

Chemicals and apparatus:

- Anhydrous sodium carbonate ✖ i
- Deionised (or distilled) water
- Balance (accurate to 0.01 g)
- Clock glass
- Beaker (250 cm^3)
- Wash bottle

- Stirring rod
- Volumetric flask (250 cm^3) and stopper
- Filter funnel
- Dropping pipette
- Safety glasses

Procedure:

NB. Wear your safety glasses.

1   Using a balance, measure accurately 2.65 g of pure anhydrous sodium carbonate on a clock glass.

2   Transfer the sodium carbonate to a clean 250 cm^3 beaker.

3   Add, with stirring, about 50 cm^3 of deionised water to the beaker.

4   Use a wash bottle to rinse the clock glass with deionised water, and add the rinsings to the beaker.

5   Continue stirring the mixture with a stirring rod until the sodium carbonate has fully dissolved.

6   Wash off the solution on the stirring rod with deionised water into the beaker, using a wash bottle.

7   Pour the solution through a clean funnel into the 250 cm^3 volumetric flask.

8   Using a wash bottle, rinse out the beaker several times with deionised water, and add the rinsings through the funnel to the solution in the flask.

9   Rinse the funnel with deionised water, allowing the water to run into the flask.

10 Fill the flask to within about 1 cm of the calibration mark, and then add the water dropwise, using a dropping pipette, until the bottom of the meniscus just rests on the calibration mark.

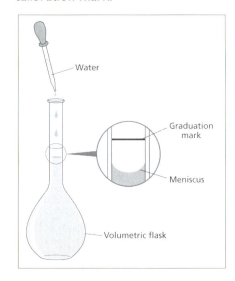

11 Stopper the flask and invert it several times to ensure an evenly mixed solution.

Conclusions/comments:

_____
_____
_____
_____

Student report, Activity 6

Title:_____**Date:**_____

Chemicals and apparatus:

_____
_____
_____

Procedure:

_____
_____
_____
_____
_____
_____
_____
_____
_____
_____
_____
_____
_____
_____
_____
_____
_____
_____
_____
_____
_____
_____
_____

Conclusions/comments:

_____

_____

_____

_____

Signature: _____

Student report, Activity 6 – diagram

1   Why is deionised water, and not tap water, used in this experiment?

_____

_____

2   The solution that you have made up in this experiment is a standard solution. Explain what is meant by a standard solution.

_____

_____

_____

3   Anhydrous sodium carbonate is a <u>primary standard</u>. Explain the underlined term.

_____

_____

4   In this experiment, state how you would ensure that no material is lost when it is being transferred from the clock glass to the beaker.

_____

_____

_____

5   In this experiment, state two steps that are taken to ensure that no material is lost when it is being transferred from the beaker to the volumetric flask.

_____

_____

_____

6   What is done at the end of the experiment to make sure that the solution is uniform? Why is this particularly important?

_____

_____

7   If you were using a 100 cm³ volumetric flask when making up a 0.1 M solution of sodium carbonate, what mass of anhydrous sodium carbonate would you measure out?

_____

Activity 7 (mandatory experiment) – Standardisation of a hydrochloric acid solution using a standard solution of sodium carbonate

Theory:

The equation for the reaction between hydrochloric acid and sodium carbonate is:

$$2HCl_{(aq)} + Na_2CO_{3(aq)} \rightarrow 2NaCl_{(aq)} + H_2O_{(l)} + CO_{2(g)}$$

The concentration of a hydrochloric acid solution may be found experimentally by titrating it with a standard solution of sodium carbonate. In the titration, a measured volume of the hydrochloric acid solution is added from a burette to a definite known volume of the solution of sodium carbonate in a conical flask, until the reaction just reaches completion. The end-point, when neutralisation just occurs, is detected using methyl orange indicator. At the end-point the indicator changes colour from yellow to peach/pink.

The titration results are used to calculate the concentration of the hydrochloric acid solution.

Chemicals and apparatus:

- 0.1 M sodium carbonate solution
- Solution of hydrochloric acid
- Methyl orange indicator 🔥
- Deionised (or distilled) water
- Burette (50 cm³)
- Retort stand
- Clamp
- Filter funnel

- Conical flask (250 cm³)
- Pipette (25 cm³)
- Pipette filler
- Wash bottle
- White tile
- White card
- Beakers (250 cm³)
- Safety glasses

Procedure:

NB. Wear your safety glasses.

1 Rinse the conical flask with deionised water.

2 Rinse the pipette with deionised water, and then with sodium carbonate solution.

3 Using the pipette, place 25 cm³ of the sodium carbonate solution in the conical flask.

4 Add 3 drops of methyl orange indicator.

5 Pour about 50 cm³ of the hydrochloric acid solution into a clean dry beaker.

6 Rinse the burette with deionised water, and clamp it vertically using the retort stand.

7 Using a funnel, rinse the burette with the hydrochloric acid solution.

8 Fill the burette with hydrochloric acid solution above the zero mark. Remove the funnel.

9 Hold a piece of white card behind the burette in order to see the level of liquid more clearly.

Allow the acid to flow from the burette into a beaker until the level of liquid is at the zero mark.

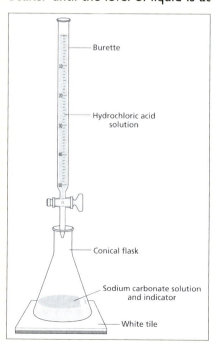

11 Carry out a rough titration by adding hydrochloric acid solution from the burette in approximately 1 cm³ lots to the conical flask, swirling the flask constantly.

12 Use the wash bottle occasionally during the titration to wash down the sides of the conical flask with deionised water.

13 Continue until the colour of the solution in the conical flask changes.

14 Note the burette reading.

15 Repeat the titration more accurately until two readings agree within 0.1 cm³.

Results:

	Values (specify units)
Rough titre	
Second titre	
Third titre	
Average of accurate titres	

Calculations:

Volume of sodium carbonate solution used in each titration	
Concentration of sodium carbonate solution	
Average volume of hydrochloric acid solution	
Concentration of hydrochloric acid solution	

Conclusions/comments:

_____

_____

_____

_____

Title:_____**Date:**_____

Chemicals and apparatus:

_____

_____

_____

Procedure:

_____

_____

_____

_____

_____

_____

_____

_____

_____

_____

_____

_____

_____

_____

_____

_____

_____

_____

_____

_____

_____

_____

_____

_____

_____

_____

_____

	Values (specify units)
Rough titre	
Second titre	
Third titre	
Average of accurate titres	

Calculations:

Volume of sodium carbonate solution used in each titration	
Concentration of sodium carbonate solution	
Average volume of hydrochloric acid solution	
Concentration of hydrochloric acid solution	

Conclusions/comments:

_____

_____

_____

_____

Signature:_____

Student report, Activity 7 – diagram

1   Why is it necessary to standardise the hydrochloric acid solution?

_____

_____

2   What is different about the procedure used in rinsing the conical flask prior to a titration compared to that used when rinsing the burette or the pipette?

_____

_____

_____

3   Why is a conical flask, rather than a beaker, used during the titrations?

_____

_____

4   Why is the funnel removed from the burette after adding the hydrochloric acid solution?

_____

_____

5   In using the burette in this experiment, why is it important (a) to clamp it vertically, and (b) to make sure that the part below the tap is full of the acid solution?

(a) _____

(b) _____

6   What steps are taken to ensure that the pipette delivers the correct volume of sodium carbonate solution into the conical flask?

_____

_____

7   Why is a rough titration carried out?

_____

_____

8   Why is an indicator necessary in this experiment?

_____

_____

9   State any safety precautions that you would take in this experiment.

_____

_____

Activity 7a (mandatory experiment) – A hydrochloric acid/sodium hydroxide titration and the use of this titration in making the salt sodium chloride

Theory:

The equation for the reaction between hydrochloric acid and sodium hydroxide is:

$$HCl_{(aq)} + NaOH_{(aq)} \rightarrow NaCl_{(aq)} + H_2O_{(l)}$$

The concentration of a sodium hydroxide solution may be found experimentally by titrating it with a standard solution of hydrochloric acid. In the titration, a measured volume of the hydrochloric acid solution is added from a burette to a definite known volume of the solution of sodium hydroxide in a conical flask, until the reaction just reaches completion. The end point, when neutralisation just occurs, is detected using methyl orange indicator. At the end-point the indicator changes colour from yellow to peach/pink.

The titration results are used to calculate the concentration of the sodium hydroxide solution. The experiment is then repeated without the indicator, and the neutral solution obtained is evaporated to give the salt (sodium chloride).

Chemicals and Apparatus:

- Sodium hydroxide solution ☒ i
- 0.1 M solution of hydrochloric acid
- Methyl orange indicator 🔥
- Deionised (or distilled) water
- Burette (50 cm³)
- Retort stand
- Clamp
- Filter funnel
- Conical flask (250 cm³)

- Pipette (25 cm³)
- Pipette filler
- Wash bottle
- White tile
- White card
- Beakers (250 cm³)
- Bunsen burner or hotplate
- Safety glasses

Procedure:

NB. Wear your safety glasses.

(a)   To find the end-point accurately

1   Rinse the conical flask with deionised water.

2   Rinse the pipette with deionised water, and then with sodium hydroxide solution.

3   Using the pipette, place 25 cm³ of the sodium hydroxide solution in the conical flask. Add 3 drops of methyl orange indicator.

4   Pour about 50 cm³ of the hydrochloric acid solution into a clean dry beaker.

5   Rinse the burette with deionised water, and clamp it vertically using the retort stand.

6   Using a funnel, rinse the burette with the hydrochloric acid solution.

7 Fill the burette with hydrochloric acid solution above the zero mark. Remove the funnel.

8 Allow the acid to flow from the burette into a beaker until the level of liquid is at the zero mark.

9 Carry out a rough titration.

10 Note the burette reading

11 Repeat the titration more accurately until two readings agree within 0.1 cm³.

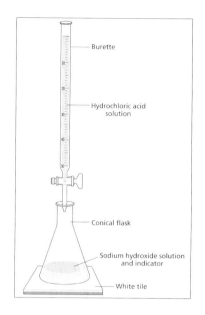

Labels: Burette; Hydrochloric acid solution; Conical flask; Sodium hydroxide solution and indicator; White tile

To obtain a sample of salt

1 Place 25 cm³ of the sodium hydroxide solution in a beaker, without any indicator.

2 Using your results from part (a) of the experiment, add just enough hydrochloric acid to exactly neutralise it.

3 Gently heat the solution until all the water has evaporated to dryness.

4 Record what happens.

Results:

	Values (specify units)
Rough titre	
Second titre	
Third titre	
Average of accurate titres	

Appearance of salt formed:

Calculations:

Volume of sodium carbonate solution used in each titration	
Average volume of hydrochloric acid solution	
Concentration of hydrochloric acid solution	
Concentration of sodium hydroxide solution	

Conclusions/comments:

Title:_____**Date:**_____

Chemicals and apparatus:

_____

_____

_____

Procedure:

_____

_____

_____

_____

_____

_____

_____

_____

_____

_____

_____

_____

_____

_____

_____

_____

_____

_____

_____

_____

_____

_____

_____

Results:

	Values (specify units)
Rough titre	
Second titre	
Third titre	
Average of accurate titres	

Appearance of salt formed:

_____

Calculations:

Volume of sodium hydroxide solution used in each titration	
Average volume of hydrochloric acid solution	
Concentration of hydrochloric acid solution	
Concentration of sodium hydroxide solution	

Conclusions/comments:

_____

_____

_____

_____

Signature: _____

Student report, Activity 7a – diagram

1   Why is it necessary to standardise the sodium hydroxide solution?

_____

_____

2   Why is the conical flask placed on a white tile during each of the titrations?

_____

_____

3   What procedure should be used to ensure that the level of solution in the burette is read correctly?

_____

_____

4   What piece of equipment is used in this experiment to measure the sodium hydroxide solution?

_____

_____

5   Why is more than one accurate titration carried out in this experiment?

_____

_____

6   What steps are taken during each of the accurate titrations in this experiment to ensure an accurate result?

_____

_____

7   How closely should the accurate titration results agree with each other?

_____

_____

8   Why is an indicator not used in the part of the experiment where the salt sample is being obtained?

_____

_____

Chapter 6

Activity 8 (mandatory experiment) – Determination of the concentration of ethanoic acid in vinegar

Theory:

Vinegar is composed mainly of water and ethanoic acid. The concentration of ethanoic acid in vinegar may be found by titrating a diluted solution of vinegar with standard sodium hydroxide solution. The equation for the titration reaction is:

$$CH_3COOH_{(aq)} + NaOH_{(aq)} \rightarrow CH_3COONa_{(aq)} + H_2O_{(l)}$$

Phenolphthalein indicator is used. At the end-point, the indicator changes colour from pink to colourless.

Chemicals and apparatus:

- 0.1 M sodium hydroxide solution ✖ i
- Phenolphthalein indicator 🔥
- Vinegar
- Deionised (or distilled) water
- Volumetric flask (250 cm³) and stopper
- Burette (50 cm³)
- Retort stand
- Clamp
- Filter funnel

- Conical flask (250 cm³)
- Pipettes (25 cm³)
- Pipette filler
- Wash bottle
- White tile
- White card
- Dropping pipette
- Beakers (250 cm³)
- Safety glasses

Procedure:

NB. Wear your safety glasses.

1. Rinse the conical flask with deionised water.
2. Rinse a pipette with deionised water, and then with sodium hydroxide solution.
3. Using the pipette, place 25 cm³ of the sodium hydroxide solution in the conical flask.
4. Add 3 drops of phenolphthalein indicator.
5. Rinse a second pipette with vinegar.
6. Using this pipette, place 25 cm³ of vinegar in a 250 cm³ volumetric flask, and dilute with water to the calibration mark.
7. Stopper the flask and invert several times to ensure an evenly-mixed solution.
8. Pour about 50 cm³ of the diluted vinegar into a clean dry beaker.
9. Rinse the burette with deionised water, and clamp it vertically using the retort stand.
10. Using a funnel, rinse the burette with the diluted vinegar.
11. Fill the burette with diluted vinegar above the zero mark. Remove the funnel.

12. Allow the diluted vinegar to flow from the burette into a beaker until the level of liquid is at the zero mark.
13. Carry out a rough titration.
14. Note the burette reading.
15. Repeat the titration more accurately until two readings agree within 0.1 cm³.

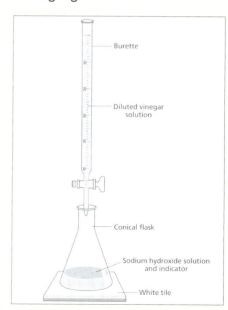

Results:

	Values (specify units)
Rough titre	
Second titre	
Third titre	
Average of accurate titres	

Calculations:

Volume of sodium hydroxide solution used in each titration	
Average volume of diluted vinegar	
Concentration of sodium hydroxide solution	
Concentration of ethanoic acid in the diluted vinegar	
Concentration of ethanoic acid in the undiluted vinegar	
Percentage (w/v) of ethanoic acid in the vinegar	

Conclusions/comments:

Title:_____ **Date:**_____

Chemicals and apparatus:

_____

_____

_____

Procedure:

_____

_____

_____

_____

_____

_____

_____

_____

_____

_____

_____

_____

_____

_____

_____

_____

_____

_____

_____

_____

Results:

	Values (specify units)
Rough titre	
Second titre	
Third titre	
Average of accurate titres	

Calculations:

Volume of sodium hydroxide solution used in each titration	
Average volume of diluted vinegar	
Concentration of sodium hydroxide solution	
Concentration of ethanoic acid in the diluted vinegar	
Concentration of ethanoic acid in the undiluted vinegar	
Percentage (w/v) of ethanoic acid in the vinegar	

Conclusions/comments:

_____

_____

_____

_____

Signature:_____

Student report, Activity 8 – diagram

1   State one safety precaution that you would take when doing this experiment.

_____

_____

2   Why is the vinegar diluted in this experiment?

_____

_____

3   What steps are taken to ensure that the pipette delivers the correct volume of vinegar into the volumetric flask?

_____

_____

4   What is the correct procedure in this experiment for bringing the solution in the volumetric flask precisely to the 250 cm³ mark?

_____

_____

5   What is the procedure for washing and filling the burette in preparation for the rough titration?

_____

_____

6   What colour change happens at the end point in each titration?

_____

_____

7   Why is phenolphthalein used as the indicator in this titration?

_____

_____

8   Why are two accurate titrations carried out in this experiment?

_____

_____

9   How closely should the accurate titration results agree with each other?

_____

Activity 9 (mandatory experiment) – Determination of the amount of water of crystallisation in hydrated sodium carbonate

Theory:

Hydrated sodium carbonate ($Na_2CO_3.xH_2O$) contains water of crystallisation. The number of molecules of water of crystallisation in the formula is x. In this experiment, the value of x in the formula is found by titration of a solution made using hydrated sodium carbonate with a standard solution of hydrochloric acid.

The equation for the reaction is:

$$2HCl_{(aq)} + Na_2CO_{3(aq)} \rightarrow 2NaCl_{(aq)} + H_2O_{(l)} + CO_{2(g)}$$

Methyl orange indicator solution is used. At the end point the indicator changes colour from yellow to peach/pink.

Chemicals and apparatus:

- 0.1 M hydrochloric acid solution
- Hydrated sodium carbonate ☒ i
- Methyl orange indicator 🔥
- Deionised (or distilled) water
- Clock glass
- Stirring rod
- Volumetric flask (250 cm³) and stopper
- Burette (50 cm³)
- Retort stand
- Clamp

- Filter funnel
- Conical flask (250 cm³)
- Pipette (25 cm³)
- Pipette filler
- Wash bottle
- White tile
- White card
- Dropping pipette
- Beakers (250 cm³)
- Safety glasses

Procedure:

NB. Wear your safety glasses.

1   Weigh accurately about 1.5 g of hydrated sodium carbonate into a beaker.

2   Add about 50 cm³ of deionised water, and stir to dissolve the sample.

3   Using a funnel, transfer all of the solution into a 250 cm³ volumetric flask.

4   Rinse the beaker with deionised water and add the washings to the volumetric flask.

5   Carefully make up the volumetric flask to the mark.

6   Stopper the flask and invert several times.

7   Rinse the conical flask with deionised water.

8   Rinse the pipette with deionised water, and then with sodium carbonate solution.

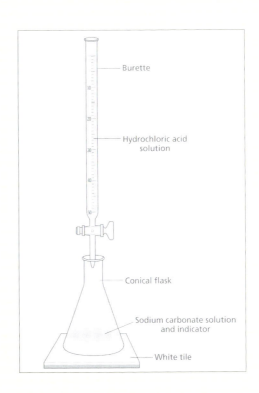

Burette

Hydrochloric acid solution

Conical flask

Sodium carbonate solution and indicator

White tile

9 Using the pipette, place 25 cm³ of the sodium carbonate solution in the conical flask. Add 3 drops of methyl orange indicator.

10 Pour about 50 cm³ of the hydrochloric acid solution into a clean dry beaker.

11 Rinse the burette with deionised water, and clamp it vertically using the retort stand.

12 Using a funnel, rinse the burette with the hydrochloric acid solution.

13 Fill the burette with hydrochloric acid solution above the zero mark. Remove the funnel.

14 Allow the hydrochloric acid solution to flow from the burette into a beaker until the level of liquid is at the zero mark.

13 Carry out a rough titration. Note the burette reading.

14 Repeat the titration more accurately until two readings agree within 0.1 cm³.

Results:

	Values (specify units)
Mass of hydrated sodium carbonate	
Rough titre	
Second titre	
Third titre	
Average of accurate titres	

Calculations:

Volume of sodium carbonate solution used in each titration	
Average volume of hydrochloric acid solution	
Concentration of hydrochloric acid solution	
Concentration of sodium carbonate solution in mol l⁻¹	
Concentration of sodium carbonate solution in g l⁻¹	
Molar mass of $Na_2CO_3.xH_2O$	
Value of x in $Na_2CO_3.xH_2O$	
Formula of hydrated sodium carbonate	
Percentage of water of crystallisation present in $Na_2CO_3.xH_2O$	

Conclusions/comments:

_____

_____

_____

_____

Title:_____**Date:**_____

Chemicals and apparatus:

_____

_____

_____

Procedure:

_____

_____

_____

_____

_____

_____

_____

_____

_____

_____

_____

_____

_____

_____

_____

_____

_____

_____

_____

_____

Results:

	Values (specify units)
Rough titre	
Second titre	
Third titre	
Average of accurate titres	

Calculations:

Volume of sodium carbonate solution used in each titration	
Average volume of hydrochloric acid solution	
Concentration of hydrochloric acid solution	
Concentration of sodium carbonate solution in mol l⁻¹	
Concentration of sodium carbonate solution in g l⁻¹	
Molar mass of Na₂CO₃.xH₂O	
Value of x in Na₂CO₃.xH₂O	
Formula of hydrated sodium carbonate	
Percentage of water of crystallisation present in Na₂CO₃.xH₂O	

Conclusions/comments:

_____

_____

_____

_____

Signature: _____

1   Describe what hydrated sodium carbonate looks like.

2   State one difference between the sodium carbonate used in this experiment and the sodium carbonate used in Experiment 5.

3   What is the correct procedure in this experiment for bringing the solution in the volumetric flask precisely to the 250cm³ mark?

4   What is done to the volumetric flask and its contents in this experiment immediately after the solution has been made up to the mark with deionised water? Why is it important to do this?

5   What is meant by water of crystallisation?

6   Why is a standard solution of hydrochloric acid used in this experiment?

7   What colour change happens at the end point in each titration?

8   In the titrations in this experiment it is preferable to use as little methyl orange as possible. What is the reason for this?

Activity 10 (mandatory experiment) – A potassium manganate(VII)/ammonium iron(II) sulfate titration

Theory:

The equation for the reaction between potassium manganate(VII) and ammonium iron(II) sulfate is:

$$MnO_4^-{}_{(aq)} + 5Fe^{2+}{}_{(aq)} + 8H^+{}_{(aq)} \rightarrow Mn^{2+}{}_{(aq)} + 5Fe^{3+}{}_{(aq)} + 4H_2O_{(l)}$$

The concentration of a potassium manganate(VII) solution may be found experimentally by titrating it with a standard solution of ammonium iron(II) sulfate. No indicator is needed, as the manganate(VII) ions are decolourised in the reaction until the end-point, when a pale pink colour persists.

The titration results are used to calculate the concentration of the potassium manganate(VII) solution.

Chemicals and apparatus:

- Potassium manganate(VII) solution
- 0.1 M ammonium iron(II) sulfate solution
- 1.5 M sulfuric acid solution ❎ i
- Deionised (or distilled) water
- Burette (50 cm³)
- Retort stand
- Clamp
- Filter funnel
- Conical flask (250 cm³)

- Pipette (25 cm³)
- Pipette filler
- Wash bottle
- White card
- White tile
- Beakers (250 cm³)
- Graduated cylinder (100 cm³)
- Safety glasses

Procedure:

NB. Wear your safety glasses.

1 Wash the pipette, burette and conical flask with deionised water.

2 Rinse the burette with the potassium manganate(VII) solution and the pipette with the iron(II) solution.

3 Place 25 cm³ of the iron(II) solution in the conical flask using the pipette. Add about 10 cm³ of dilute sulfuric acid.

4 Using a funnel, fill the burette with potassium manganate(VII) solution, making sure that the part below the tap is filled before adjusting to zero. Because of the dark colour of the potassium manganate(VII) solution, it may be necessary to take readings from the top of the meniscus. Carry out a rough titration and a number of accurate

titrations until two titres agree to within 0.1 cm³. The end-point of the titration is detected by the first persisting pale pink colour.

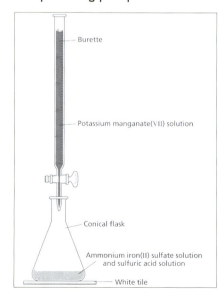

Burette

Potassium manganate(VII) solution

Conical flask

Ammonium iron(II) sulfate solution and sulfuric acid solution

White tile

Results:

	Values (specify units)
Rough titre	
Second titre	
Third titre	
Average of accurate titres	

Calculations:

Volume of iron(II) solution used in each titration	
Concentration of iron(II) solution	
Average volume of potassium manganate(VII) solution	
Concentration of potassium manganate(VII) solution	

Conclusions/comments:

_____

_____

_____

_____

Title: _____ **Date:** _____

Chemicals and apparatus:

_____
_____
_____

Procedure:

_____
_____
_____
_____
_____
_____
_____
_____
_____
_____
_____
_____
_____
_____
_____
_____
_____
_____
_____

Results:

	Values (specify units)
Rough titre	
Second titre	
Third titre	
Average of accurate titres	

Calculations:

Volume of iron(II) solution used in each titration	
Concentration of iron(II) solution	
Average volume of potassium manganate(VII) solution	
Concentration of potassium manganate(VII) solution	

Conclusions/comments:

_____

_____

_____

_____

Signature: _____

Student report, Activity 10 – diagram

1 Why is it necessary to standardise the potassium manganate(VII) solution?

_____

_____

2 Why is dilute sulfuric acid used in making up the standard solution of ammonium iron(II) sulfate?

_____

_____

3 Why is the funnel removed from the burette after adding the potassium manganate(VII) solution?

_____

_____

4 Why is dilute sulfuric acid added to the ammonium iron(II) sulfate solution in the conical flask prior to each of the titrations in this experiment?

_____

_____

5 Why are readings usually taken from the top of the meniscus in this experiment?

_____

_____

6 What steps are taken to ensure that the pipette delivers the correct volume of ammonium iron(II) sulfate solution into the conical flask?

_____

_____

7 Why is ammonium iron(II) sulfate suitable for use as a primary standard?

_____

_____

8 Why is no indicator necessary in this experiment?

_____

_____

Activity 11 (mandatory experiment) – Estimation of iron(II) in an iron tablet using a standard solution of potassium manganate(VII)

Theory:

To find the iron(II) content of an iron tablet, a known number of tablets are first dissolved in dilute sulfuric acid. This solution is then titrated against a standard solution of potassium manganate(VII).

The reaction is represented by the equation:

$$MnO_4^-{}_{(aq)} + 5Fe^{2+}{}_{(aq)} + 8H^+{}_{(aq)} \rightarrow Mn^{2+}{}_{(aq)} + 5Fe^{3+}{}_{(aq)} + 4H_2O_{(l)}$$

The titration results are used to calculate the mass of iron in an iron tablet.

Chemicals and apparatus:

- 0.005 M potassium manganate(VII) solution
- Iron tablets
- 1.5 M sulfuric acid solution ☒ i
- Deionised (or distilled) water
- Electronic balance
- Clock glass
- Dropping pipette
- Mortar and pestle
- Volumetric flask (250 cm³)
- Burette (50 cm³)
- Retort stand

- Clamp
- Filter funnel
- Conical flask (250 cm³)
- Pipette (25 cm³)
- Pipette filler
- Wash bottle
- White card
- White tile
- Beakers (250 cm³)
- Safety glasses

Procedure:

NB. Wear your safety glasses.

1 Find the mass of five iron tablets.

2 Crush the weighed tablets in a mortar and pestle.

3 Transfer all the ground material to a beaker, and dissolve it in about 100 cm³ of dilute sulfuric acid.

4 Transfer all of this solution (including washings) to a 250 cm³ volumetric flask and make the solution up to the mark with deionised water.

5 Stopper the flask and invert it several times. This is the solution containing iron(II) ions.

6 Wash the pipette, burette and conical flask with deionised water.

7 Rinse the burette with the potassium manganate(VII) solution and the pipette with the iron(II) solution.

8 Place 25 cm³ of the iron(II) solution in the conical flask using the pipette. Add about 10 cm³ of dilute sulfuric acid.

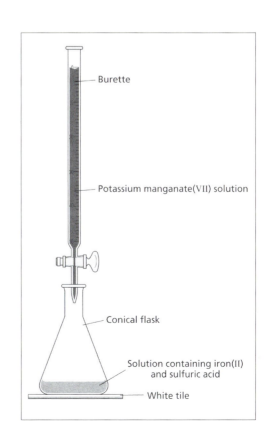

Burette

Potassium manganate(VII) solution

Conical flask

Solution containing iron(II) and sulfuric acid

White tile

9 Using a funnel, fill the burette with potassium manganate(VII) solution, making sure that the part below the tap is filled before adjusting to zero. Because of the dark colour of the potassium manganate(VII) solution, take readings from the top of the meniscus.

10 Carry out a rough titration and a number of accurate titrations until two titres agree to within 0.1 cm³. The end-point of the titration is detected by the first persisting pale pink colour.

Results:

	Values (specify units)
Mass of iron tablets	
Rough titre	
Second titre	
Third titre	
Average of accurate titres	

Calculations:

Volume of iron(II) solution used in each titration	
Average volume of potassium manganate(VII) solution	
Concentration of potassium manganate(VII) solution	
Concentration of iron(II) solution	
Volume of iron(II) solution (in total)	
Moles of iron in this volume	
Mass of iron present in the tablets	
Mass of iron present in one tablet	

Conclusions/comments:

_____

_____

_____

_____

Title:_____**Date:**_____

Chemicals and apparatus:

_____
_____
_____

Procedure:

_____
_____
_____
_____
_____
_____
_____
_____
_____
_____
_____
_____
_____
_____
_____
_____
_____
_____
_____
_____

Results:

	Values (specify units)
Mass of iron tablets	
Rough titre	
Second titre	
Third titre	
Average of accurate titres	

Calculations:

Volume of iron(II) solution used in each titration	
Average volume of potassium manganate(VII) solution	
Concentration of potassium manganate(VII) solution	
Concentration of iron(II) solution	
Volume of iron(II) solution (in total)	
Moles of iron in this volume	
Mass of iron present in the tablets	
Mass of iron present in one tablet	

Conclusions/comments:

_____

_____

_____

_____

Signature: _____

Student report, Activity 11 – diagram

1 Why is a standard solution of potassium manganate(VII) needed in this experiment?

_____

_____

2 Why did you use five iron tablets in this experiment?

_____

_____

3 Why is dilute sulfuric acid used in making up the solution of iron tablets?

_____

_____

_____

4 Using your titration results, calculate the percentage by mass of iron(II) in the iron tablets that you used.

_____

_____

5 Why is dilute sulfuric acid added to the iron(II) solution in the conical flask prior to each of the titrations in this experiment?

_____

_____

6 How is the endpoint detected in this experiment?

_____

_____

7 What would you observe if insufficient sulfuric acid were used during the titrations in this experiment? Explain why this happens.

_____

_____

_____

8 If drops of potassium manganate(VII) solution form high up on the sides of the conical flask during one of your titrations, what action would you take to deal with this problem?

_____

_____

9 Prior to the rough titration, what steps are taken to minimise error?

_____

_____

_____

10 State one safety precaution that you would take in this experiment.

_____

Chapter 7

Theory:

Iodine reacts with sodium thiosulfate according to the equation:

$$I_{2(aq)} + 2S_2O_3^{2-}{}_{(aq)} \rightarrow 2I^-{}_{(aq)} + S_4O_6^{2-}{}_{(aq)}$$

The concentration of a sodium thiosulfate solution may be found experimentally by titrating it with a standard solution of iodine.

In an iodine-thiosulfate titration, the iodine solution (0.06 M) is formed in the conical flask by the reaction of a 0.02 M solution of potassium iodate with excess potassium iodide. The thiosulfate solution is then added from the burette until the colour of the mixture in the flask changes from brown to pale yellow. A freshly prepared solution of starch is now added. The starch acts as an indicator. When it is added, it forms a blue-black colour with iodine. This colour disappears at the end-point.

The titration results are used to calculate the concentration of the sodium thiosulfate solution.

Chemicals and apparatus:

- 0.02 M potassium iodate solution 🔥
- 0.5 M potassium iodide solution
- Dilute (1 M) sulfuric acid ❌ i
- Sodium thiosulfate solution
- Starch indicator solution
- Deionised (or distilled) water
- Burette (50 cm³)
- Retort stand
- Clamp
- Filter funnel

- Conical flask (250 cm³)
- Pipette (25 cm³)
- Pipette filler
- Wash bottle
- White card
- White tile
- Beakers (250 cm³)
- Graduated cylinders (100 cm³)
- Safety glasses

Procedure:

NB. Wear your safety glasses.

1. Wash the pipette, burette and conical flask with deionised water. Rinse the pipette with potassium iodate solution and the burette with the sodium thiosulfate solution.

2. Place 25 cm³ of the potassium iodate solution in the conical flask using the pipette.

3. Using graduated cylinders, add 20 cm³ of dilute sulfuric acid, followed by 10 cm³ of 0.5 M potassium iodide solution.

4. Using a funnel, fill the burette with sodium thiosulfate solution, making sure that the part below the tap is filled before adjusting to zero.

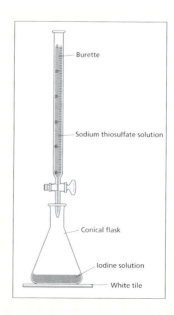

Burette

Sodium thiosulfate solution

Conical flask

Iodine solution

White tile

5 Add the sodium thiosulfate solution from the burette to the flask. Swirl the flask continuously and occasionally wash down the walls of the flask with deionised water using a wash bottle.

6 Add a few drops of the starch indicator solution just prior to the end-point, when the colour of the solution fades to pale yellow. Upon addition of the indicator a blue-black colour should appear.

7 The end-point of the titration is detected by a colour change from blue-black to colourless. Note the burette reading.

8 Carry out a number of accurate titrations, adding the sodium thiosulfate solution dropwise approaching the end-point, until two titres agree to within 0.1 cm^3.

Results:

	Values (specify units)
Rough titre	
Second titre	
Third titre	
Average of accurate titres	

Calculations:

Volume of iodine solution used in each titration	
Concentration of iodine solution	
Average volume of sodium thiosulfate solution	
Concentration of sodium thiosulfate solution	

Conclusions/comments:

_____

_____

_____

_____

Title:_____**Date:**_____

Chemicals and apparatus:

_____
_____
_____

Procedure:

_____
_____
_____
_____
_____
_____
_____
_____
_____
_____
_____
_____
_____
_____
_____
_____
_____
_____
_____
_____
_____
_____
_____

Results:

	Values (specify units)
Rough titre	
Second titre	
Third titre	
Average of accurate titres	

Calculations:

Volume of iodine solution used in each titration	
Concentration of iodine solution	
Average volume of sodium thiosulfate solution	
Concentration of sodium thiosulfate solution	

Conclusions/comments:

_____

_____

_____

_____

Signature: _____

Student report, Activity 12 – diagram

1 Why is a standard solution of iodine needed in this experiment?

_____

_____

2 What steps are taken to ensure that the pipette delivers the correct volume of potassium iodate solution into the volumetric flask?

_____

_____

3 What is the procedure for washing and filling the burette in preparation for the rough titration?

_____

_____

_____

4 Why are two accurate titrations carried out in this experiment?

_____

_____

5 When excess potassium iodide solution is added to the mixture of potassium iodate solution and dilute sulfuric acid in the conical flask, iodine is formed. Why is excess potassium iodide solution used?

_____

_____

6 How is the endpoint detected in this experiment?

_____

_____

7 State one safety precaution that you would take in this experiment.

_____

8 During each titration, the walls of the conical flask are occasionally washed down with deionised water. Why is this done?

_____

_____

9 How closely should the accurate titration results agree with each other?

_____

10 What colour change is observed when potassium iodide solution is added to the mixture of potassium iodate solution and dilute sulfuric acid in the conical flask?

_____

Activity 13 (mandatory experiment)
Determination of the percentage (w/v) of hypochlorite in bleach

Theory:

Many commercial bleaches are simply solutions of hypochlorite salts such as sodium hypochlorite (NaOCl) or calcium hypochlorite ($Ca(OCl)_2$). To determine the amount of hypochlorite in bleach, the bleach is first diluted. Hypochlorite ion in the diluted bleach is then reacted with excess iodide ion in the presence of acid to generate an iodine solution:

$$ClO^-_{(aq)} + 2I^-_{(aq)} + 2H^+_{(aq)} \rightarrow Cl^-_{(aq)} + I_{2(aq)} + H_2O_{(l)}$$

The iodine solution produced is titrated against sodium thiosulfate solution using starch solution as indicator. Iodine reacts with sodium thiosulfate according to the equation:

$$I_{2(aq)} + 2S_2O_3^{2-}{}_{(aq)} \rightarrow 2I^-_{(aq)} + S_4O_6^{2-}{}_{(aq)}$$

The titration results are used to calculate the percentage of hypochlorite in the bleach.

Chemicals and apparatus:

- Bleach solution ✖ i
- Potassium iodide
- Dilute sulfuric acid ✖ i
- 0.1 M sodium thiosulfate solution
- Starch indicator solution
- Deionised (or distilled) water
- Volumetric flask (250 cm³) and stopper
- Burette (50 cm³)
- Retort stand
- Clamp
- Filter funnel

- Conical flask (250 cm³)
- Pipette (25 cm³)
- Pipette filler
- Wash bottle
- White card
- White tile
- Beakers (250 cm³)
- Graduated cylinders (100 cm³)
- Dropping pipette
- Safety glasses

Procedure:

NB. Wear your safety glasses.

1 Using a pipette, add 25 cm³ of bleach to a 250 cm³ volumetric flask, and make the solution up to the mark with deionised water.

2 Stopper the flask and invert it several times.

3 Wash the pipette, burette and conical flask with deionised water.

4 Rinse the pipette with the diluted bleach solution and the burette with the sodium thiosulfate solution.

5 Place 25 cm³ of the diluted bleach solution in the conical flask using the pipette.

6 Add 10 cm³ of dilute sulfuric acid and 1 g of potassium iodide to the conical flask – an iodine solution will be formed.

7 Using a funnel, fill the burette with sodium thiosulfate solution, making sure that the part below the tap is filled before adjusting to zero.

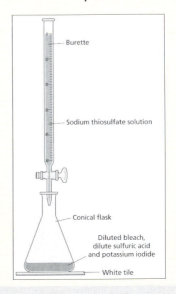

Burette

Sodium thiosulfate solution

Conical flask

Diluted bleach, dilute sulfuric acid and potassium iodide

White tile

8 Carry out a rough titration and a number of accurate titrations until two titres agree to within 0.1 cm³. Each time, add a few drops of the starch indicator solution just prior to the end-point, when the colour of the solution fades to **pale** yellow. The end-point of the titration is detected by a colour change from blue-black to colourless.

Results:

	Values (specify units)
Rough titre	
Second titre	
Third titre	
Average of accurate titres	

Calculations:

Volume of iodine solution used in each titration	
Average volume of sodium thiosulfate solution	
Concentration of sodium thiosulfate solution	
Concentration of iodine solution	
Concentration of hypochlorite in the diluted bleach solution	
Concentration of hypochlorite in the bleach	
Percentage (w/v) of hypochlorite in the bleach	

Conclusions/comments:

_____

_____

_____

_____

Title:_____**Date:**_____

Chemicals and apparatus:

_____
_____
_____

Procedure:

_____
_____
_____
_____
_____
_____
_____
_____
_____
_____
_____
_____
_____
_____
_____
_____
_____
_____
_____
_____
_____

Results:

	Values (specify units)
Rough titre	
Second titre	
Third titre	
Average of accurate titres	

Calculations:

Volume of iodine solution used in each titration	
Average volume of sodium thiosulfate solution	
Concentration of sodium thiosulfate solution	
Concentration of iodine solution	
Concentration of hypochlorite in the diluted bleach solution	
Concentration of hypochlorite in the bleach	
Percentage (w/v) of hypochlorite in the bleach	

Conclusions/comments:

_____

_____

_____

_____

Signature: _____

1 Why is a standard solution of sodium thiosulfate needed in this experiment?

_____

_____

2 Why is the bleach diluted before being placed in the conical flask?

_____

_____

3 What colour change is observed when potassium iodide and dilute sulfuric acid solution are added to the diluted bleach in the conical flask? Explain why this colour change occurs.

_____

_____

_____

4 The potassium iodide has two functions in this experiment. What are they?

_____

_____

5 What colour change is observed in the conical flask as sodium thiosulfate solution is added from the burette?

_____

6 When is the indicator added to the reaction mixture in the conical flask?

_____

_____

7 State one safety precaution that you would take in this experiment.

_____

8 What colour change occurs when the indicator is added to the contents of the conical flask? Explain why this colour change occurs.

_____

_____

9 What colour change takes place at the endpoint? Explain why this colour change occurs.

_____

_____

Activity 14 (mandatory experiment) – Determination of the heat of reaction of hydrochloric acid with sodium hydroxide

Theory:

The heat of reaction, ΔH, of a chemical reaction is the heat in kilojoules released or absorbed when the numbers of moles of reactants indicated, in the balanced equation describing the reaction, react completely. The heat change depends on the amounts of reactants involved, and this is indicated by the balanced equation.

In this experiment a definite volume of a standard hydrochloric acid solution is mixed with an equal volume of sodium hydroxide solution of the same concentration in an insulated container. The rise in temperature of the mixture is then measured and from this the heat of reaction is calculated.

The equation for the reaction is:

$$HCl_{(aq)} + NaOH_{(aq)} \rightarrow NaCl_{(aq)} + H_2O_{(l)}$$

Chemicals and apparatus:

- 1 M hydrochloric acid ✖ i
- 1 M sodium hydroxide
- Thermometers

- Polystyrene cups
- Graduated cylinders
- Safety glasses

Procedure:

NB. Wear your safety glasses.

1 Place 50 cm³ of the 1 M hydrochloric acid solution into one of the polystyrene cups.

2 Place 50 cm³ of the 1 M sodium hydroxide solution into the second polystyrene cup.

3 Measure the temperature of each solution.

4 Quickly add the base to the acid, stirring with a thermometer.

5 Record the highest temperature reached.

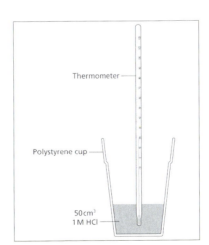

Thermometer

Polystyrene cup

50 cm³
1 M HCl

Results:

	Values (specify units)
Volume of hydrochloric acid solution reacted	
Volume of sodium hydroxide solution reacted	
Concentration of hydrochloric acid solution	
Concentration of sodium hydroxide solution	
Initial temperature of hydrochloric acid solution	
Initial temperature of sodium hydroxide solution	
Average initial temperature	
Highest temperature reached	
Rise in temperature	
Rise in temperature in kelvins	

Calculations:

Number of moles of hydrochloric acid	
Total volume of reaction mixture	
Mass of reaction mixture in kg	
Specific heat capacity of water	4.2 kJkg^{-1}K^{-1}
Heat change (mcΔT)	
Heat of reaction	

Conclusions/comments:

_____

_____

_____

_____

Title:_____**Date:**_____

Chemicals and apparatus:

_____

_____

_____

Procedure:

_____

_____

_____

_____

_____

_____

_____

_____

Results:

	Values (specify units)
Volume of hydrochloric acid solution reacted	
Volume of sodium hydroxide solution reacted	
Concentration of hydrochloric acid solution	
Concentration of sodium hydroxide solution	
Initial temperature of hydrochloric acid solution	
Initial temperature of sodium hydroxide solution	
Average initial temperature	
Highest temperature reached	
Rise in temperature	
Rise in temperature in kelvins	

Calculations:

Number of moles of hydrochloric acid used	
Total volume of reaction mixture	
Mass of reaction mixture in kg	
Specific heat capacity of water	4.2 kJkg⁻¹K⁻¹
Heat change (mcΔT)	
Heat of reaction	

Conclusions/comments:

_____

_____

_____

_____

Signature: _____

Student report, Activity 14 - diagram

1   State one safety precaution that you would take in this experiment.

2   Why is it important to prevent heat loss in this experiment?

3   What precautions are taken to prevent heat loss in this experiment?

4   Suggest an extra precaution that could be taken to prevent heat loss.

5   Why is it important to stir with the thermometer immediately after mixing the acid and the base?

6   Textbooks suggest that the heat of reaction for this experiment should have been -57 kJmol-1. If your result was different to this, suggest reasons for the difference.

7   Define heat of reaction.

8   What is the sign of ΔH for an exothermic reaction?

9   What assumption is made about the density of the reaction mixture in this experiment when doing the calculations?

10   When 100 cm³ of 1.0 M sodium hydroxide reacted with excess hydrochloric acid solution, 5.72 kJ of heat were produced. What is the heat of reaction (ΔH) for the reaction between sodium hydroxide and hydrochloric acid?

11 An experiment was carried out to measure the heat of reaction (ΔH) for the reaction between nitric acid and potassium hydroxide, according to the equation $HNO_{3(aq)} + KOH_{(aq)} \rightarrow KNO_{3(aq)} + H_2O_{(l)}$. Using an insulated container, 100 cm³ of each of the reactant 1 M solutions were mixed and the rise in temperature was found to be 6.85 kelvins. What is the heat of reaction (ΔH) for the reaction between potassium hydroxide and nitric acid?

Activity 15 (mandatory experiment) – Monitoring the rate of production of oxygen from hydrogen peroxide, using manganese dioxide as a catalyst

Theory:

Hydrogen peroxide decomposes into water and oxygen as follows:

$$H_2O_{2(l)} \rightarrow H_2O_{(l)} + 1/2\ O_{2(g)}$$

In the absence of a catalyst, this reaction occurs much too slowly to be monitored. However, manganese dioxide acts as a suitable catalyst, and, when it is added, the reaction occurs at a measurable rate.

Method 1 – using an inverted graduated cylinder

Chemicals and apparatus:

- Hydrogen peroxide (20 volumes)
- Powdered manganese(IV) oxide ✖ n
- 100 cm³ graduated cylinder
- Beehive shelf
- Large trough
- Conical flask with suitable stopper and delivery

- tube
- Stop-clock
- Small test tube
- Thread
- Teat pipette
- Safety glasses

Procedure:

NB. Wear your safety glasses.

1 Measure out 5 cm³ of hydrogen peroxide.

2 Dilute to 50 cm³ with water.

3 Place this solution in the conical flask.

4 Weigh about 0.5 g manganese(IV) oxide into the small test tube.

5 Suspend the test tube in the conical flask using the thread. (The manganese(IV) oxide and the hydrogen peroxide should not come into contact until the stop-clock is started.)

6 Fill the graduated cylinder with water from the trough, and invert it onto the beehive shelf.

7 Assemble the apparatus for the collection, by displacement of water, of the oxygen produced.

8 Bring the manganese(IV) oxide into contact with the hydrogen peroxide by loosening the stopper sufficiently to allow the thread to fall into the flask.

9 Shake vigorously, starting the stop-clock as the manganese(IV) oxide comes into contact with the hydrogen peroxide solution.

10 Record the total volume of gas in the graduated cylinder every 30 seconds.

11 Draw a graph of total volume of oxygen against time, putting time on the horizontal axis.

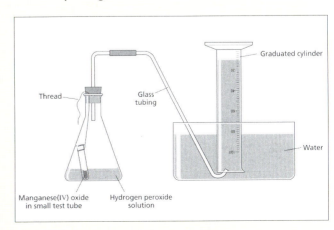

Chemicals and apparatus:

- Hydrogen peroxide (20 volumes)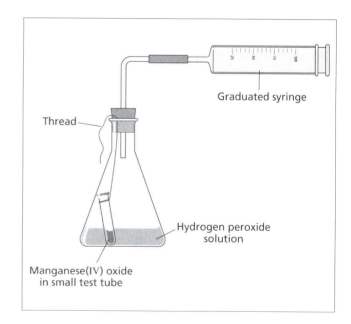
- Powdered manganese(IV) oxide ✖ n
- 100 cm³ gas syringe
- Conical flask with single-holed stopper and suitable delivery tube
- Stop-clock
- Small test tube
- Thread
- Graduated cylinder
- Safety glasses

Procedure:

NB. Wear your safety glasses.

1 Measure out 8 cm³ of hydrogen peroxide.

2 Dilute to 50 cm³ with water.

3 Place this solution in the conical flask.

4 Weigh about 0.5 g manganese(IV) oxide into the small test tube.

5 Suspend the test tube in the conical flask using the thread. (The manganese(IV) oxide and the hydrogen peroxide should not come into contact until the stop-clock is started.)

6 Connect the delivery tube to the gas syringe, ensuring that the syringe is set to the zero reading.

7 Bring the manganese(IV) oxide into contact with the hydrogen peroxide by loosening the stopper sufficiently to allow the thread to fall into the flask.

8 Shake vigorously, starting the stop-clock as the manganese(IV) oxide comes into contact with the hydrogen peroxide solution.

9 Record the total volume of gas in the syringe every 30 seconds.

10 Using the graph paper provided, draw a graph of total volume of oxygen against time, putting time on the horizontal axis.

Results:

Time (min)	Total volume of oxygen (cm³)

Conclusions/comments:

Title:_____**Date:**_____

Chemicals and apparatus:

_____
_____
_____

Procedure:

_____
_____
_____
_____
_____
_____
_____
_____
_____
_____
_____
_____
_____
_____
_____
_____
_____
_____
_____
_____
_____
_____
_____
_____
_____

Results:

Time (min)	Total volume of oxygen (cm^3)

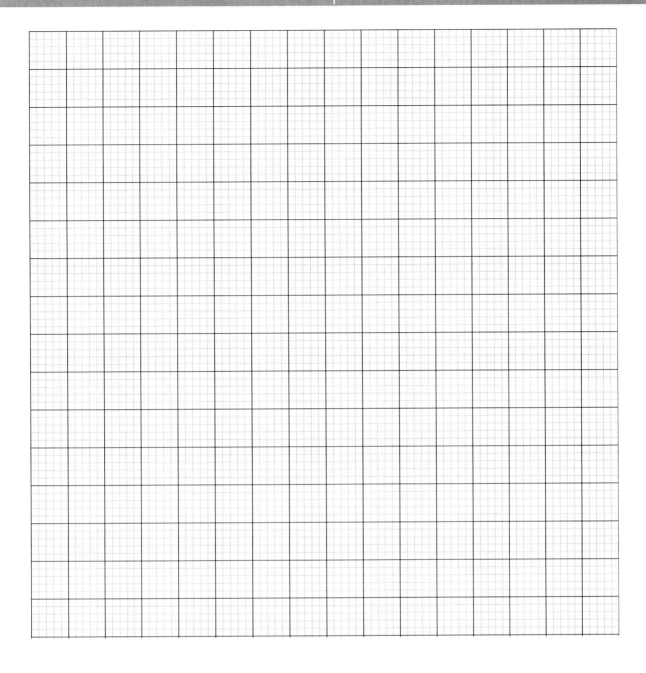

Conclusions/comments:

_____

_____

_____

_____

Signature:_____

1   Explain why the manganese dioxide should not come in contact with the hydrogen peroxide until the stop-clock is started.

_____

_____

2   Describe the shape of the graph, referring to the rate of reaction.

_____

_____

3   Why is the slope of the graph steepest at the beginning?

_____

4   How can you tell when the reaction has finished?

_____

5   Would increasing the amount of catalyst make any difference to the rate?

_____

6   Would increasing the concentration of hydrogen peroxide solution make any difference to the shape of the graph? Explain.

_____

_____

7   How might the use of a very concentrated hydrogen peroxide solution make the experiment impractical?

_____

8   Use your graph to estimate the volume of oxygen that had been collected after 135 seconds.

_____

9   Use your graph to estimate the number of seconds that had elapsed when 25 cm^3 of oxygen gas had been collected.

_____

10   Use your graph to calculate the instantaneous rate of reaction after two minutes. Record your result here.

_____

Activity 16 (mandatory experiment) – Studying the effects on the reaction rate of (i) concentration and (ii) temperature, using sodium thiosulfate solution and hydrochloric acid

Theory:

Sodium thiosulfate solution reacts with hydrochloric acid solution according to the following equation:

$$2HCl_{(aq)} + Na_2S_2O_{3(aq)} \rightarrow 2NaCl_{(aq)} + SO_{2(aq)} + S_{(s)} + H_2O_{(l)}$$

The pale yellow precipitate of sulfur formed gradually obscures a cross marked on paper placed beneath the reaction flask. The time taken to obscure the cross, which is inversely proportional to the rate of reaction, depends on variables such as temperature and concentration. By varying one of these and keeping other variables constant, the effect on rate can be studied.

The inverse of the time taken to obscure the cross is the measure of reaction rate used in this experiment.

(a) Effect of concentration

Temperature is kept constant while concentration is varied.

Chemicals and apparatus:
- 0.1 M sodium thiosulfate solution
- Dilute hydrochloric acid (3 M) **✗** i
- Conical flasks
- Graduated cylinders
- Stop-clock
- Safety glasses

Procedure:

NB. Wear your safety glasses.

1. Place 100 cm³ of the sodium thiosulfate solution into a conical flask.
2. Add 10 cm³ of 3 M hydrochloric acid to the flask and swirl, while starting the stop clock at the same time.
3. Place the flask on a piece of white paper marked with a cross.
4. Stop the clock when the cross disappears completely.
5. Record the time taken.
6. Repeat the experiment using 80 cm³ of sodium thiosulfate solution mixed with water to make the volume up to 100 cm³.
7. Repeat using 60, 40 and 20 cm³ of sodium thiosulfate solution respectively in turn, made up

to 100 cm³ with water in each case. (If the initial sodium thiosulfate concentration is 0.1 M, subsequent concentrations will be 0.08 M, 0.06 M, 0.04 M and 0.02 M respectively.)

8. Record the results in the table.
9. Draw a graph of reaction rate i.e. 1/time (vertical axis), against concentration of thiosulfate solution (horizontal axis).

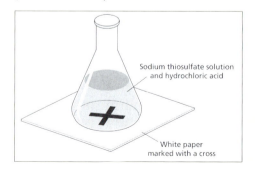

Sodium thiosulfate solution and hydrochloric acid

White paper marked with a cross

Results:

Concentration of thiosulfate	Reaction time (sec)	Rate of reaction (sec⁻¹)
0.1 M		
0.08 M		
0.06 M		
0.04 M		
0.02 M		

Conclusions/comments:

_____

_____

_____

_____

(b) Effect of temperature

Concentration is kept constant while temperature is varied.

Chemicals and apparatus:

- 0.05 M sodium thiosulfate solution
- Dilute hydrochloric acid (3 M) ❌ i
- Conical flasks
- Graduated cylinder
- Stop-clock

- Thermometer or temperature probe
- Bunsen burner
- Tripod
- Wire gauze
- Safety glasses

Procedure:

NB. Wear your safety glasses.

1 Place 100 cm³ of 0.05 M sodium thiosulfate solution into a conical flask at room temperature.

2 Add 5 cm³ of 3 M hydrochloric acid, starting a stop clock at the same time.

3 Swirl the flask immediately and place it on a piece of white paper marked with a cross.

4 Record the exact temperature of the contents of the flask.

5 Record the time taken for the cross to disappear.

6 Repeat the experiment, heating the thiosulfate solution to temperatures of approximately 30 °C, 40 °C, 50 °C and 60 °C respectively (before adding the hydrochloric acid, swirling and taking the exact temperature).

7 Record the results in the table.

8 Draw a graph of reaction rate i.e. 1/time (vertical axis) against temperature of reaction mixture (horizontal axis).

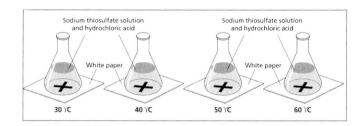

Results:

Temperature	Reaction time (sec)	Rate of reaction (sec⁻¹)

Conclusions/comments:

_____

_____

_____

_____

(a) Effect of concentration

Title:_____**Date:**_____

Chemicals and apparatus:

_____
_____
_____

Procedure:

_____
_____
_____
_____
_____
_____
_____
_____
_____
_____
_____
_____
_____
_____
_____
_____
_____
_____
_____
_____
_____

Results:

Concentration of thiosulfate	Reaction time (sec)	Rate of reaction (sec⁻¹)
0.1 M		
0.08 M		
0.06 M		
0.04 M		
0.02 M		

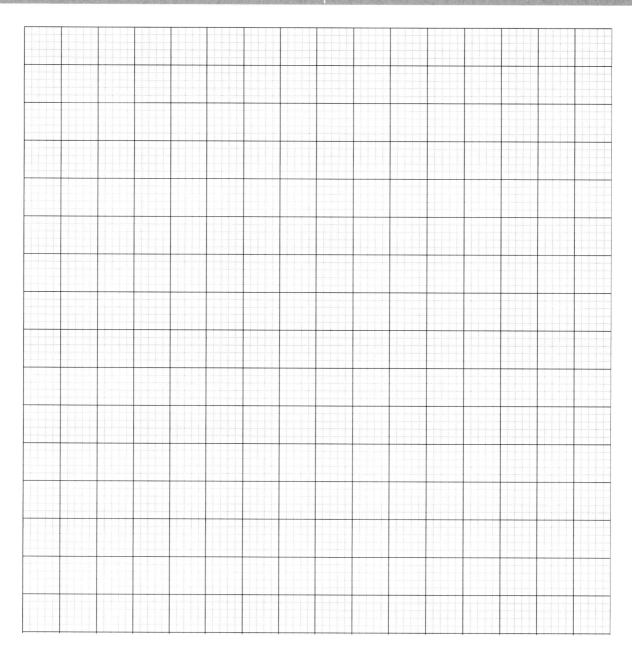

Conclusions/comments:

_____

_____

_____

_____

Signature:_____

Student report, Activity 16 – diagram

(a) Effect of concentration

(b) Effect of temperature

Title:_____**Date:**_____

Chemicals and apparatus:

_____
_____
_____

Procedure:

_____
_____
_____
_____
_____
_____
_____
_____
_____
_____
_____
_____
_____
_____
_____
_____
_____

Results:

Temperature	Reaction time (sec)	Rate of reaction (sec^{-1})

Graph

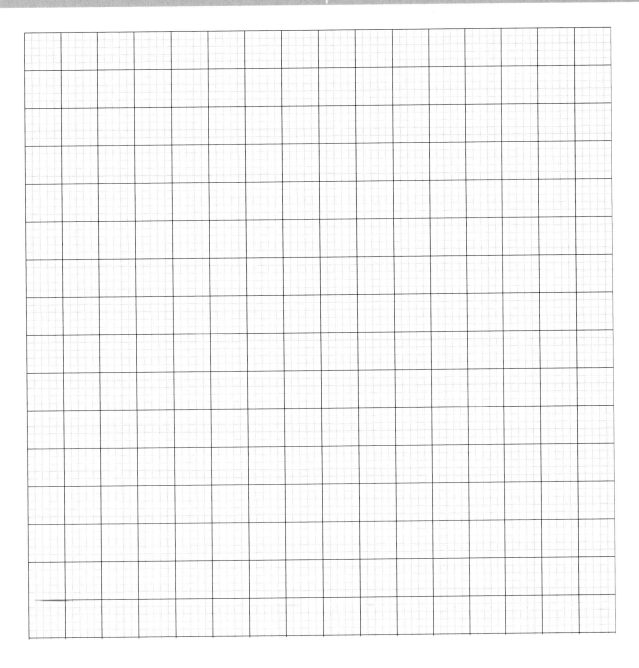

Conclusions/comments:

_____

_____

_____

_____

Signature: _____

(b) Effect of temperature

1   What is the effect on reaction time of increasing the concentration of sodium thiosulfate solution?

_____

2   What is the effect on reaction rate of increasing the concentration of hydrochloric acid solution?

_____

3   What is the effect on reaction time of increasing the temperature of sodium thiosulfate solution?

_____

4   What is the effect on reaction rate of increasing the temperature of hydrochloric acid solution?

_____

5   Name two factors that affect the rate of chemical reactions.

_____

6   Suggest a reason why it would be very difficult to carry out experiment (b) at temperatures above 60°C.

_____

7   Name the yellow precipitate that obscures the cross beneath the reaction flask.

_____

8   State one safety precaution that you would take in this experiment.

_____

Activity 17 (mandatory experiment) –
The extraction of clove oil from cloves by steam distillation

Theory:

A natural product is any chemical produced in nature, either by plants or animals. One of the techniques used to extract natural products from plants is steam distillation. This technique is used in this experiment to separate clove oil from cloves. Steam distillation allows substances to be distilled that, if heated on their own to higher temperatures, might partially decompose.

Chemicals and apparatus:

- 5–10 g whole cloves
- Anti-bumping granules
- PVC gloves
- Steam generator
- Quickfit apparatus for distillation with steam inlet
- Bunsen burner

- Tripod
- Wire-gauze
- Retort stands and clamps
- Water-bath
- Safety glasses

Procedure:

NB. Wear your safety glasses.

1   Set up the Quickfit apparatus as in the figure. Ensure that the assembled apparatus has a safety opening to the atmosphere as in the diagram.
2   Weigh out about 5–10 grams of whole cloves.
3   Place them in the pear-shaped flask.
4   Cover with a little warm water.
5   Connect the steam generator to the rest of the apparatus.
6   Heat the steam generator to boiling point.
7   Regulate the heat so that a constant supply of steam is supplied to the pear shaped flask.
8   Monitor the level of boiling water in the steam generator during the experiment. (If it falls too low, remove the heat, carefully loosen the safety valve, and top up the steam generator with hot water. Reconnect everything and resume heating.)

9   Collect the distillate, which should have a pale milky appearance.
10 Note the strong smell of clove oil.
11 When sufficient product has been collected, disconnect the steam generator to prevent suck-back, and turn off the heat.

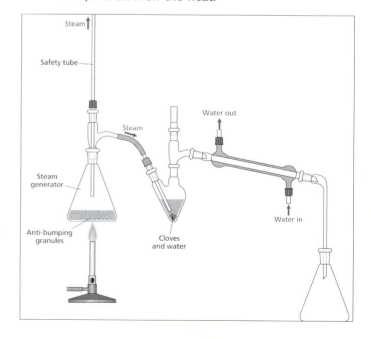

Results:

_____

_____

_____

_____

Conclusions/comments:

_____

_____

_____

_____

Title:_____**Date:**_____

Chemicals and apparatus:

_____
_____
_____

Procedure:

_____
_____
_____
_____
_____
_____
_____
_____
_____
_____
_____
_____
_____
_____
_____
_____
_____
_____
_____
_____
_____
_____
_____
_____
_____

Results:

_____
_____
_____
_____

Conclusions/comments:

_____

_____

_____

_____

Signature:_____

1   Why is a safety opening in the apparatus essential?

2   What is the appearance of the distillate?

3   What is the odour of the distillate?

4   Why are anti-bumping granules used in the steam generator?

5   What is the function of the condenser in this experiment?

6   Why is it important to disconnect the steam generator when the product has been collected?

7   If cloves were not available, suggest an alternative natural product that might be used.

8   Explain why clove oil cannot be distilled directly from cloves.

9   Name a technique that could be used to completely separate the clove oil from water after the distillate from this experiment has been collected.

10   Name four everyday examples of organic natural products.

Activity 18 (mandatory experiment) – Preparation of soap

Theory:

Soaps are the sodium or potassium salts of long-chain carboxylic acids. To make soap in this experiment, a mixture of vegetable oils and potassium hydroxide is reacted.

Vegetable oils are esters of long-chain carboxylic acids and of the alcohol propane-1,2,3-triol (glycerol). The base hydrolysis of these substances produces glycerol, and the salt of the acid present, i.e. soap. For example, if the animal fat contains esters of stearic acid, the reaction is as follows:

$$
\begin{array}{lll}
C_{17}H_{35}COO - CH_2 & C_{17}H_{35}COOK & CH_2OH \\
\quad\quad\quad | & + & | \\
C_{17}H_{35}COO - CH \;\; + 3KOH \rightarrow & C_{17}H_{35}COOK \quad\quad + & CHOH \\
\quad\quad\quad | & + & | \\
C_{17}H_{35}COO - CH_2 & C_{17}H_{35}COOK & CH_2OH \\
& \text{potassium stearate} & \text{glycerol}
\end{array}
$$

Potassium stearate is the soap formed in this particular case. The soap is a long-chain hydrocarbon with an ionic group at the end. The non-polar part dissolves grease while the ionic end dissolves in water. This combination of properties gives it its cleansing action.

Chemicals and apparatus:

- Vegetable oil
- Potassium hydroxide
- Ethanol
- Saturated sodium chloride solution (brine)
- Condenser
- Flask
- Still head
- Thermometer (-10 → 110 °c)

- Receiver with side arm
- Filter funnel
- Retort stands and clamps
- Anti-bumping granules
- Bunsen burner
- Filter paper
- Water bath
- Safety glasses

Procedure:

NB. Wear your safety glasses.

1   Add 3 cm³ of vegetable oil, 2.5 g of potassium hydroxide and 20 cm³ ethanol, along with a few anti-bumping granules, to the flask.

2   Swirl to allow proper mixing.

3   Assemble the apparatus for reflux, greasing all joints in the process.

4   Reflux the mixture for 20 minutes, using a water bath.

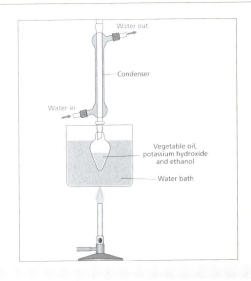

5 Allow the apparatus to cool, and reassemble for distillation.

6 Remove the ethanol by distillation.

7 Add about 15 cm³ of hot water to dissolve the residue.

8 Pour this solution into a beaker of brine. The soap will precipitate out.

9 Filter the soap.

10 Test the soap for its lathering qualities by shaking a small sample of it with water. There may be traces of potassium hydroxide still present, so contact with skin should be kept to a minimum.

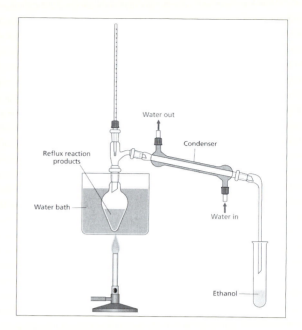

Results:

_____

_____

_____

_____

Conclusions/comments:

_____

_____

_____

_____

Title:_____**Date:**_____

Chemicals and apparatus:

_____
_____
_____
_____
_____
_____
_____
_____
_____
_____
_____
_____
_____
_____
_____
_____
_____
_____
_____
_____
_____
_____

Procedure:

_____
_____
_____
_____
_____
_____
_____
_____

Results:

_____
_____
_____
_____

Chapter 11

Conclusions/comments:

_____

_____

_____

_____

Signature:_____

Student report, Activity 18 – diagram

1   What is the function of the ethanol in this experiment?

_____

2   Why are anti-bumping granules added to the reaction mixture?

_____

3   What is meant by refluxing?

_____

_____

4   Why is the mixture refluxed?

_____

5   Would boiling the reaction mixture in an open beaker be as successful as refluxing? Explain your answer.

_____

_____

6   How is the ethanol removed at the end of the reflux procedure?

_____

7   Why is it desirable to remove the ethanol after reflux?

_____

8   Why is a minimum of hot water used to dissolve the residue from the distillation?

_____

9   How would you prepare a beaker of brine?

_____

_____

10   Why is the residue placed in a beaker of brine?

_____

11   How does potassium stearate act as soap?

_____

_____

Activity 19 (mandatory experiment) – Preparation and properties of ethyne

Theory:

Calcium dicarbide reacts with water producing ethyne and calcium hydroxide:

$$CaC_{2(s)} + 2H_2O_{(l)} \rightarrow C_2H_{2(g)} + Ca(OH)_{2(aq)}$$

The ethyne gas formed in the reaction is insoluble in water, and is therefore collected by downward displacement of water. Impurities present in the gas such as hydrogen sulfide and phosphine cause an unpleasant odour. They can be removed by bubbling the gas through acidified copper sulfate solution. The impurities are caused by the hydrolysis of traces of calcium sulfide and calcium phosphide present in the calcium dicarbide.

Chemicals and apparatus:

- Calcium dicarbide
- Water
- Acidified copper(II) sulfate solution   n
- Limewater   i
- Very dilute acidified potassium manganate(VII) solution   i
- Very dilute bromine water
- Dropping funnel
- Buchner flask and one-hole stopper

- Bottle and two-holed stopper
- Teat-pipettes
- Test tubes
- Solid stoppers
- Delivery tubes
- Trough
- Beehive shelf
- Retort stands and clamps
- Safety glasses

Procedure:

NB. Wear your safety glasses.

Preparation:

1. Place a few pieces of calcium dicarbide in a Buchner flask.
2. Set up the apparatus as in the diagram.
3. Add water from the dropping funnel, a few drops at a time.
4. Collect the gas produced in test tubes by displacement of water.
5. Stopper each gas-filled test tube under water.
6. Discard the first test tube filled, as it contains a mixture of air and ethyne.

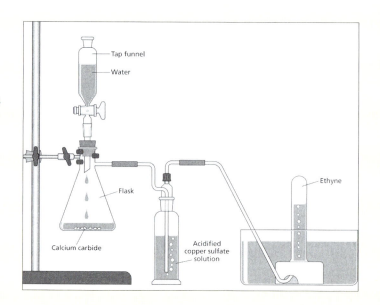

Investigation of Properties:

1. Ignite a test tube of the gas.
2. Describe the flame.
3. Add a few drops of limewater to the test-tube, stopper and shake well.
4. Record what happens.
5. Add a few drops of a very dilute solution of bromine water to a test tube of gas, stopper quickly and shake well.
6. Record what you see.
7. Add a few drops of very dilute acidified potassium manganate solution to a test tube of gas, stopper quickly and shake well.
8. Record what you see.

Results:

Test	Observation	Explanation
Gas sample ignited		
Limewater added after combustion		
Bromine water added		
Acidified potassium manganate solution added		

Conclusions and comments:

_____

_____

_____

_____

Title:_____**Date:**_____

Chemicals and apparatus:

_____

_____

_____

Procedure:

_____

_____

_____

_____

_____

_____

_____

_____

_____

_____

_____

_____

_____

_____

_____

_____

_____

_____

Risk management:

_____

_____

_____

Results:

Test	Observation	Explanation
Gas sample ignited		
Limewater added after combustion		
Bromine water added		
Acidified potassium manganate solution added		

Conclusions/comments:

_____

_____

_____

_____

Signature:_____

Student report, Activity 19 – diagram

1 Describe the appearance of calcium carbide.

2 Could ethyne gas be collected by displacement of methylbenzene instead of by displacement of water? Explain your answer.

3 Name two impurities that are likely to be formed in the preparation of ethyne gas.

4 How can impurities be removed during the preparation of ethyne gas?

5 Why is the first test tube of collected gas discarded?

6 Why is a sooty flame produced when a sample of ethyne gas is burned?

7 Which combustion product of ethyne is responsible for turning limewater milky?

8 Write a balanced equation for the complete combustion of ethyne in oxygen.

9 Describe a test for unsaturation.

10 In testing for unsaturation, why is it desirable to use very dilute solutions?

Activity 20 (mandatory experiment) – Preparation and properties of ethene

Theory:

Ethanol is dehydrated, using hot aluminium oxide as a catalyst, to produce ethene:

$$\underset{\text{heat}}{\overset{Al_2O_3}{C_2H_5OH \rightarrow C_2H_4 + H_2O}}$$

The ethene gas formed in the reaction is insoluble in water, and is therefore collected by downward displacement of water.

Chemicals and apparatus:

- Ethanol 🔥
- Aluminium oxide powder
- Limewater ✖ i
- Very dilute bromine water ☠ 🧪
- Very dilute acidified potassium manganate(vii) solution ✖ i
- Glass wool
- Large heat-resistant glass boiling tube with one-holed rubber stopper

- Test tubes and stoppers
- Trough
- Delivery tube
- Bunsen burner
- Retort stand and clamp
- Safety glasses

Procedure:

NB. Wear your safety glasses.

Preparation:

1 Pour ethanol into the boiling tube to a depth of about 2 cm.

2 Push in enough glass wool to soak up all of the ethanol.

3 Set up the apparatus as shown in the diagram, with about 2 g of aluminium oxide heaped halfway along the boiling tube.

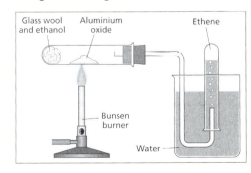

4 Heat the catalyst strongly, and occasionally heat the ethanol gently to drive the vapour over the catalyst.

5 Collect a few test tubes of ethene by displacement of water, stoppering the test tubes when they are filled. The first test tube filled can be discarded, as it contains a mixture of air and ethene.

6 When the reaction has concluded, remove the tube from the water.

7 Turn off the Bunsen burner.

Investigation of properties:

1. Ignite the ethene gas in one of the test tubes.
2. Record whether the flame is coloured or clear, smoky or clean.
3. Pour a few drops of limewater into the test tube.
4. Stopper, shake well and record what you see.
5. Add a few drops of very dilute bromine water to another test tube of ethene gas.
6. Stopper, shake well and record what you see.
7. Add a few drops of very dilute acidified potassium manganate(VII) solution to a third test tube of the ethene gas.
8. Stopper, shake well and record what you see.

Results:

Test	Observation	Explanation
Sample ignited		
Limewater added after combustion		
Bromine water added		
Acidified potassium manganate solution added		

Conclusions and comments:

_____

_____

_____

_____

Title:_____**Date:**_____

Chemicals and apparatus:

_____

_____

_____

Procedure:

_____

_____

_____

_____

_____

_____

_____

_____

_____

_____

_____

_____

_____

_____

_____

_____

_____

_____

_____

_____

_____

_____

_____

_____

_____

_____

_____

Results:

Test	Observation	Explanation
Sample ignited		
Limewater added after combustion		
Bromine water added		
Acidified potassium manganate solution added		

Conclusions/comments:

_____

_____

_____

_____

Signature: _____

Student Report, Activity 20 – diagram

1   What is the function of the glass wool in this experiment?

_____

2   Why is the glass wool inserted after the ethanol has been added to the test tube?

_____

3   What colour is aluminium oxide?

_____

4   State one safety precaution you would take when doing this experiment.

_____

5   What steps are taken to avoid suck-back? Why are such steps necessary?

_____

_____

6   Why is the first test tube of gas collected discarded?

_____

7   What colour flame results when the gas is burned?

_____

8   Why is limewater added to the tube after combustion?

_____

9   What is observed on adding limewater? Explain this result.

_____

_____

10   Is ethene saturated or unsaturated? Justify your answer.

_____

_____

Activity 21 (mandatory experiment) – Preparation and properties of ethanal

Theory:

A primary alcohol such as ethanol can be oxidised to an aldehyde such as ethanal using sodium dichromate solution, acidified with sulfuric acid. By using more ethanol than oxidising agent, and by distilling off the ethanal as soon as it is formed, further oxidation to ethanoic acid can be minimized. In this experiment, ethanol is oxidized to ethanal:

$$C_2H_5OH \xrightarrow{Na_2Cr_2O_7/H_2SO_4} CH_3CHO$$

The balanced equation for the reaction is:

$$3C_2H_5OH + Cr_2O_7^{2-} + 8H^+ \rightarrow 3CH_3CHO + 2Cr^{3+} + 7H_2O$$

Chemicals and apparatus:

- Dilute sulfuric acid ✖ i
- Sodium dichromate ☠ 🔥
- Concentrated sulfuric acid 🧪
- Anti-bumping granules 🔥
- Ethanol 🔥
- Ethanal 🔥 ✖ i
- Potassium manganate(VII) solution
- Fehling's solution no. 1 ✖ n
- Fehling's solution no. 2 🧪
- Silver nitrate solution ✖ i
- Sodium hydroxide solution 🧪
- Aqueous ammonia solution ✖ i
- Dilute nitric acid 🧪
- Ice

- PVC gloves
- Quickfit apparatus for distillation
- Dropping funnel
- Plastic filter funnel
- Retort stands
- Clamps
- Conical flask
- Beakers
- Heating mantle or Bunsen/tripod/wire gauze
- Test tubes
- Water bath
- Graduated cylinder
- Thermometer
- Safety glasses

Procedure:

NB. Wear your safety glasses.

Preparation:

1   Set up the apparatus for distillation with addition, as shown in the diagram.

2   Place 12 cm³ of water in the reaction flask.

3   Add some anti-bumping granules to the flask.

4   Slowly, with swirling and cooling under a cold-water tap, add 4 cm³ of concentrated sulfuric acid.

5   Dissolve 10 g of sodium dichromate in 10 cm³ of water in a clean small beaker.

6   Add 10 cm³ of ethanol to the beaker and stir.

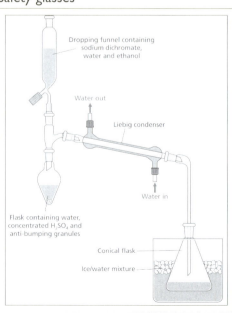

Dropping funnel containing sodium dichromate, water and ethanol

Water out

Liebig condenser

Water in

Flask containing water, concentrated H₂SO₄ and anti-bumping granules

Conical flask

Ice/water mixture

7 Place this solution in the dropping funnel, using a small plastic funnel.

8 Arrange the conical collection flask so that it is standing in a large beaker of ice/water.

9 Heat the dilute acid gently until it just boils, and then remove the heat source.

10 Slowly add the alcohol/dichromate mixture from the dropping funnel so as to maintain a gentle boiling.

11 The ethanal distils off as it is formed.

12 Redistill the impure ethanal and collect the fraction boiling between 20°C and 23°C.

Properties:

(a) Oxidation by acidified potassium manganate(VII) solution

1 To about 2 cm³ of ethanal in a test-tube, add 1 cm³ of potassium manganate(VII) solution and 4 cm³ of dilute sulfuric acid.

2 Warm the test tube in a water-bath and shake gently. Record any colour change that occurs.

(b) Oxidation by Fehling's solution

1 Mix about 1 cm³ each of Fehling's solution no.1 and Fehling's solution no.2 in a test tube.

2 Swirl the contents so that the blue precipitate initially formed will dissolve.

3 Add 1 cm³ of ethanal, heat gently and shake.

4 Record any change that occurs.

(c) Silver mirror test (oxidation by ammoniacal silver nitrate)

1 Place 3 cm³ of silver nitrate solution and 1 cm³ of sodium hydroxide solution in a **clean** test tube.

2 Add aqueous ammonia solution drop wise, with shaking, until the precipitate formed in step 1 is just dissolved.

3 Add two or three drops of ethanal.

4 Shake, and warm in a water bath.

5 Record any change that occurs.

6 Rinse out the test tube with dilute nitric acid and then with water.

Results:

Test	Observation	Explanation
Acidified potassium manganate(VII) added		
Fehling's solution added		
Tollen's reagent added		

Conclusions and comments:

_____

_____

_____

_____

Title:_____**Date:**_____

Chemicals and apparatus:

_____
_____
_____

Procedure:

_____
_____
_____
_____
_____
_____
_____
_____
_____
_____
_____
_____
_____
_____
_____
_____
_____
_____
_____
_____
_____
_____
_____
_____
_____
_____
_____
_____
_____
_____

Results:

Test	Observation	Explanation
Acidified potassium manganate(VII) added		
Fehling's solution added		
Tollen's reagent added		

Conclusions/comments:

_____

_____

_____

_____

Signature:_____

Student report, Activity 21 – diagram

1   Why is cooling necessary when adding concentrated sulfuric acid to water?

_____

2   Why is the collection vessel in this experiment surrounded by ice/water?

_____

3   What colour is sodium dichromate?

_____

4   What is likely to happen if excess sodium dichromate were used by mistake?

_____

5   Why is heating not required once addition of the alcohol/dichromate mixture has begun?

_____

6   Why is it important to distil off the ethanal as it is produced?

_____

7   What is the colour of the mixture in the reaction flask at the end of the reaction?

_____

8   Apart from a colour change, what else is observed in the flask as the ethanol/dichromate mixture was added to the hot sulfuric acid?

_____

9   What colour change is observed when ethanal reacts with acidified potassium manganate(VII) solution? What does this colour change indicate?

_____

_____

10   What is formed when ethanal reacts with Fehling's solution? What does this result indicate?

_____

_____

11   In the silver mirror test, why is it essential to use a very clean test tube?

_____

_____

12   What does the formation of a silver mirror indicate?

_____

13   Fill in the blanks in the following passage:

The reactions referred to in questions 6, 8 and 10 above are all .... reactions. A positive result in each of them indicates that the reactant is an .... When a .... is added to Fehling's solution or Tollens' reagent, no colour change is noted. It can be concluded that .... are not easily ....

Theory:

A primary alcohol such as ethanol can be oxidised to a carboxylic acid such as ethanoic acid by refluxing with sodium dichromate solution, acidified with sulfuric acid. During refluxing, any ethanal or ethanol vapour that leaves the reaction mixture condenses and falls back into the reaction flask for further oxidation. By using more oxidizing agent than alcohol, and by refluxing the reaction mixture for 20 to 30 minutes, any aldehyde formed is oxidized to carboxylic acid. In this experiment, ethanol is oxidized to ethanoic acid:

$$C_2H_5OH \xrightarrow{Na_2Cr_2O_7/H_2SO_4} CH_3COOH$$

The balanced equation for the reaction is:

$$3C_2H_5OH + 2Cr_2O_7{}^{2-} + 16H^+ \rightarrow 3CH_3COOH + 4Cr^{3+} + 11\ H_2O$$

Chemicals and apparatus:

- Dilute sulfuric acid ✖ i
- Sodium dichromate ☠ 🔥
- Concentrated sulfuric acid 🧪
- Anti-bumping granules
- Ethanol 🔥
- Deionised water
- Universal indicator paper
- Magnesium ribbon 🔥
- Anhydrous sodium carbonate ✖ i
- Safety glasses
- PVC gloves

- Quickfit apparatus for distillation
- Dropping funnel
- Retort stands
- Clamps
- Water bath
- Ice bath
- Graduated cylinder
- Thermometer
- Bunsen burner
- Tripod
- Wire gauze

Procedure:

NB. Wear your safety glasses.

Preparation:

1. Place 10 cm³ of dilute sulfuric acid in the reaction flask.
2. Add some anti-bumping granules to the flask.
3. Add in 9 g of sodium dichromate and dissolve in the acid by careful swirling.
4. Add 6 cm³ of concentrated sulfuric acid slowly, with swirling and cooling, in an ice bath.

5   Set up the apparatus for reflux with addition, as shown in the diagram.

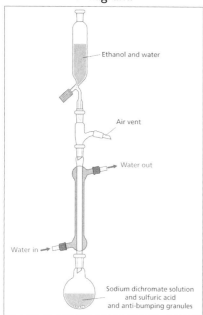

6   Mix 2 cm³ of ethanol and 10 cm³ of deionised water.

7   Place this mixture in the dropping funnel.

8   Very slowly add the solution from the dropping funnel down the condenser, while swirling and cooling the contents of the flask.

9   Remove the dropping funnel and still head from the top of the assembled apparatus.

10   Using a water-bath, boil the mixture gently for about half an hour.

11   When the apparatus is cool enough to handle safely, dismantle and rearrange for distillation as in the second diagram.

12   Heat directly without a water bath, as the boiling point of the mixture will eventually exceed 100 °C.

13   Distil off about 15 cm³. This is aqueous ethanoic acid.

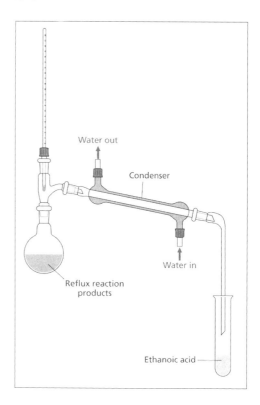

Properties:

1   Smell the reaction product by carefully wafting some of the vapour towards your nose.

2   Compare its smell to that of ethanol, and of vinegar. Record your observations.

3   Dip some universal indicator paper in the distillate to measure the pH. Record your result.

4   Drop a 5 cm clean strip of magnesium into some distillate in a test tube, and swirl. Record your observations.

5   Add 1 g of anhydrous sodium carbonate powder to some distillate in a test tube, and swirl. Record your observations.

6   Carefully add 2 drops of concentrated sulfuric acid to some distillate in a test tube. Add 1 cm³ of ethanol and warm gently.

7   Carefully smell the reaction product. Record your observations.

Results:

Test	Observation	Explanation
Odour detected		
Universal indicator paper dipped	pH =	
Magnesium strip added		
Anhydrous sodium Carbonate added		
Ethanol with concentrated sulfuric acid mixed		

Conclusions and comments:

_____

_____

_____

_____

Title: _____ **Date:** _____

Chemicals and apparatus:

_____
_____
_____

Procedure:

_____
_____
_____
_____
_____
_____
_____
_____
_____
_____
_____
_____
_____
_____
_____
_____
_____
_____
_____
_____
_____
_____
_____
_____
_____
_____
_____
_____
_____
_____
_____
_____
_____
_____
_____

Results:

Test	Observation	Explanation
Odour detected		
Universal indicator paper dipped pH =		
Magnesium strip added		
Anhydrous sodium Carbonate added		
Ethanol with concentrated sulfuric acid mixed		

Conclusions/comments:

_____

_____

_____

_____

Signature:_____

Student report, Activity 22 – diagram

1   Why are anti-bumping granules added to the reaction flask?

2   Why is the reaction flask cooled by ice/water as concentrated sulfuric acid is being added?

3   State one safety precaution that you would take in this experiment.

4   Why are the dropping funnel and still head removed before refluxing begins?

5   Why is refluxing necessary?

6   Why is a water bath used for heating during refluxing?

7   What is the colour of the reaction mixture after refluxing?

8   Why is direct heating rather than a water bath used to distil off the product?

9   What everyday substance has an odour that resembles the odour of the product?

10   In what way does the universal indicator paper show that the product is acidic?

11   How could you show that carbon dioxide gas is produced when sodium carbonate is added to the product?

12   Describe the smell and appearance of the substance produced when the product is heated with ethanol and concentrated sulfuric acid.

CHROMATOGRAPHY

Activity 23 (mandatory experiment) – Separation of a mixture of indicators using paper chromatography or thin-layer chromatography or column chromatography

Theory:

Chromatography is a type of separation technique that involves the use of a mobile phase and a stationary phase to separate the components of a mixture from each other. The separation occurs because different components of a mixture are attracted to different extents to the mobile phase as compared to the stationary phase. A component that is attracted less to the stationary phase will move more quickly during the separation than one that is attracted less to the mobile phase.

Separation of a mixture of indicators using paper chromatography

Theory:

In paper chromatography, a small spot of the mixture to be separated is placed on a rectangle of chromatography paper, on a line near the lower edge of the rectangle. This end of the rectangle is then dipped into a pool of a suitable solvent, care being taken that the spot of mixture is above the surface of the solvent. The solvent rises up the paper. The separation is carried out in a closed container, to avoid loss of solvent by evaporation.

Results obtained using paper chromatography can be recorded in terms of R_f values. The R_f value for a component of a mixture is equal to:

$$\frac{\text{Distance travelled by the component}}{\text{Distance travelled by the solvent front}}$$

R_f values can be used to tentatively identify the components of a mixture of indicators.

Chemicals and apparatus:
- Universal indicator solution 🔥 (or other mixture of indicators)
- Methyl orange solution 🔥
- Phenolphthalein solution 🔥
- Dilute ammonia solution ⚗️

- Water/ethanol/ammonia solvent (5:2:1)
- Capillary tubes
- Paper chromatography tank
- Safety glasses

Procedure:

NB. Wear your safety glasses.

1   Add the solvent to the bottom of the tank to a depth of about 10 mm.

2   Cover the tank, and allow to stand for a few hours. This will allow the tank to become saturated with solvent vapour.

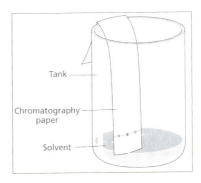

Tank

Chromatography paper

Solvent

3   Make a line with a pencil about 3 cm from the bottom of a rectangular sheet of chromatography paper, and another line near the top.

4   Using a capillary tube, place a small spot of each indicator and of the mixture of indicators at different points on the line near the bottom of the paper.

5   Dry the spots using a hair drier and repeat.

6   Place the chromatogram in the tank, taking care that the solvent level in the tank is below the line on which the indicator samples are spotted.

7   Cover the tank.

8   Run the chromatogram, until the solvent reaches the line near the top of the paper.

9   Remove and dry.

10 Calculate and record the R_f values of each indicator. (Ammonia vapour is used to locate phenolphthalein.)

Results:

Distance travelled by solvent front	
Distance travelled by methyl orange	
R_f value	
Distance travelled by phenolphthalein	
R_f value	
Distance travelled by fastest moving component of mixture	
R_f value	
Distance travelled by next fastest moving component of mixture	
R_f value	
Distance travelled by next fastest moving component of mixture	
R_f value	
Distance travelled by next fastest moving component of mixture	
R_f value	

Conclusions and comments:

_____

_____

_____

_____

Theory:

For thin layer chromatography, thin layer chromatography plates and a glass tank are needed. The thin layer chromatography tank contains a suitable solvent, and should be saturated with solvent vapour before use.

A small spot of the mixture to be separated is placed on a line near the edge of the plate. Another line is drawn near the opposite edge of the plate. The plate is then placed in the tank, and care is taken to ensure that the spot of mixture is above the surface of the solvent. The solvent travels rapidly up the plate. Results obtained using thin layer chromatography can be recorded in terms of R_f values. The R_f value for a component of a mixture is equal to:

$$\frac{\text{Distance travelled by the component}}{\text{Distance travelled by the solvent front}}$$

R_f values can be used to tentatively identify the components of a mixture of indicators.

Chemicals and apparatus:

- Universal indicator solution 🔥
- Methyl orange solution 🔥
- Phenolphthalein solution 🔥
- Industrial methylated spirits 🔥
- Dilute ammonia solution
- Capillary tubes

- Thin-layer chromatography tank
- Filter paper
- Thin layer plates
- Hair drier
- Safety glasses

Procedure:

NB. Wear your safety glasses.

1   Cut a piece of filter paper to fit around the walls of the tank.
2   Add enough industrial methylated spirits to the tank to allow it to saturate the filter paper and give a depth of about 10 mm at the bottom.
3   Cover the tank, and allow to stand for 10 minutes.
4   Using a capillary tube, place a small spot of each sample on a line drawn a little more than 10 mm from the bottom of the plate.
5   Allow to dry – a hair drier may be used to speed up drying.
6   Draw a horizontal line near the top of the plate.
7   Stand the plate carefully in the tank, making sure that the samples are above the surface of the liquid.
8   Cover the tank, and allow the solvent front to rise up the plate to the line near the top.
9   Remove the plate.
10 Allow to dry.
11 Note the colours of the zones.
12 Calculate and record the R_f values of each indicator. (Ammonia vapour is used to locate phenolphthalein.)

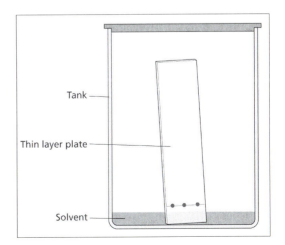

Results:

Distance travelled by solvent front	
Distance travelled by methyl orange	
R_f value	
Distance travelled by phenolphthalein	
R_f value	
Distance travelled by fastest moving component of mixture	
R_f value	
Distance travelled by next fastest moving component of mixture	
R_f value	
Distance travelled by next fastest moving component of mixture	
R_f value	
Distance travelled by next fastest moving component of mixture	
R_f value	

Conclusions and comments:

Theory:

In column chromatography, a small amount of the mixture to be separated is placed on a suitable column, and a suitable solvent is used to carry the mixture through the column. A particularly useful type of column is a solid phase extraction column (see diagram). Solid phase extraction columns vary in size and polarity. There is a certain amount of trial and error involved in selecting a suitable solvent and column for the separation of the components of a particular mixture.

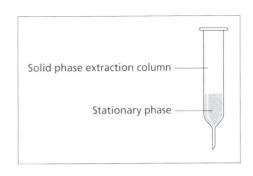

Solid phase extraction columns can be used to separate the components of a mixture of indicators.

Chemicals and apparatus:

- Universal indicator solution 🔥
- Methyl orange solution 🔥
- Phenolphthalein solution 🔥
- Methanol 🔥 ☠
- 35 % methanol solution
- 60 % methanol solution

- Dilute sodium hydroxide solution
- Plastic disposable syringe
- Solid phase extraction column
- Adaptor (to connect syringe to column)
- Small test tubes and test tube rack
- Safety glasses

Procedure:

NB. Wear your safety glasses and gloves.

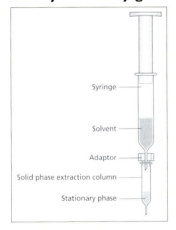

1 Using the plunger, flush the solid phase extraction column through with methanol, and then with water a number of times.

2 Place a sample of the mixture of indicators on the top of the column. The sample should cover the top of the column to a depth of slightly less than half the length of the column solid phase.

3 Half fill the syringe with air and, using the adaptor, attach the syringe to the column.

4 Using the plunger, gently force the mixture into the column.

5 Add about 4 cm³ of 35 % methanol solution to the syringe.

6 Using the adaptor, attach the syringe to the column.

7 Have several small test tubes available to collect the different components of the mixture.

8 Using the plunger force the mixture through the column.

9 Collect the different components of the mixture in separate test tubes.

10 When no more coloured liquid is emerging from the column, repeat steps 6 to 9, using 60 % methanol instead of 35 % methanol.

11 Add a little dilute sodium hydroxide solution to each of the test tubes collected.

12 Wash out the column alternately with 100 % methanol followed by excess water until no colour remains in the column.

Results:

Component	Colour	Colour with NaOH solution	Indicator identified
1			
2			
3			
4			
5			
6			

Conclusions and comments:

_____

_____

_____

_____

Student report, Activity 23

Title:_____**Date:**_____

Chemicals and apparatus:

_____
_____
_____

Procedure:

_____
_____
_____
_____
_____
_____
_____
_____
_____
_____
_____
_____
_____
_____
_____
_____
_____
_____
_____
_____
_____
_____

Results:

Conclusions/comments:

_____

_____

_____

_____

Signature: _____

Student report, Activity 23 – diagram

Activity 23 Questions

1   In paper and thin layer chromatography, why is the tank not used immediately after the chromatography solvent has been added?

2   Why are two lines usually drawn on a paper or thin layer chromatogram?

3   In general, when is it possible to separate components of a mixture using paper chromatography (or using thin layer chromatography)?

4   When two substances are found to have different R_f values, what does this mean?

5   When is it possible to separate components of a mixture using column chromatography?

6   What is the purpose of the syringe when components of a mixture are being separated using a solid phase extraction column?

7   Why is it necessary to flush a solid phase extraction column with methanol and then with water before using it to carry out a separation?

8   Why is filter paper placed around the walls of a thin layer chromatography tank?

9   Why is an adaptor necessary when using a solid phase extraction column?

10   What was the (1) mobile phase (2) stationary phase in your chromatography experiment?

Activity 24 (mandatory experiment) – Simple experiments to illustrate Le Chatelier's Principle

(a) The equilibrium between $CoCl_4^{2-}$ and $Co(H_2O)_6^{2+}$

Theory:

Le Chatelier's Principle states that when a system at equilibrium is disturbed, the equilibrium shifts in such a way as to minimize the effect of the disturbance. Reversible reactions involving cobalt chloride are suitable for illustrating Le Chatelier's Principle because they involve very definite colour changes.

When cobalt chloride is dissolved in water, the compound dissociates into its ions, Cl^- and Co^{2+}, and then $Co(H_2O)_6^{2+}$, which is pink in aqueous solution, is formed. A reversible reaction then occurs, in which water and $CoCl_4^-$, which is blue, are formed.

The following equilibrium is established:

$$CoCl_4^{2-}{}_{(aq)} + 6H_2O_{(l)} \rightleftharpoons Co(H_2O)_6^{2+}{}_{(aq)} + 4Cl^-{}_{(aq)}$$

Blue   Pink

The forward reaction is exothermic. The equilibrium between the two species can be disturbed (i) by adding Cl^- ions or water or (ii) by changing the temperature. In both cases the changes that occur are as predicted by Le Chatelier's Principle.

This experiment is used to demonstrate the effects of **concentration changes** and of **temperature changes** on an equilibrium mixture.

Chemicals and apparatus:

- Deionised water
- Concentrated hydrochloric acid
- Cobalt(II) chloride
- Crushed ice
- Test tubes
- Boiling tubes and racks

- Dropping pipettes
- Pyrex beakers (250 cm³)
- Measuring cylinders (100 cm³)
- electronic balance
- Safety glasses

Procedure:
NB. Wear your safety glasses.

(i)   **Preparing the solution for the experiment**

1   Dissolve 4 g of cobalt chloride-6-water in 40 cm³ of deionised water. A pink solution should be formed, indicating that the equilibrium lies on the right hand side.

2   Keep a 2 cm³ sample of this solution in a test tube for reference.

3   In a fume cupboard, add concentrated hydrochloric acid, with stirring, until a violet solution is formed.

4   Keep a 2 cm³ sample of this solution in another test tube for reference.

5   Add more concentrated hydrochloric acid – this

will produce a blue colour, while adding water will restore the pink colour. By trial and error produce an 'in between' violet (or lilac) colour which will contain the two cobalt ions.

6   Place the violet solution in each of six boiling tubes to a depth of about 2 cm.

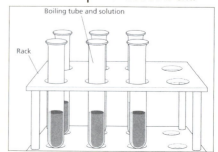

(ii) **To study the effects of concentration changes on the equilibrium**

1. Keep one boiling tube as a control.
2. Add water to a second tube using a dropping pipette. Record what happens.
3. In a fume cupboard, add concentrated hydrochloric acid to a third tube using a dropping pipette. Record what happens.

(iii) **To study the effects of temperature changes on the equilibrium**

1. Keep one boiling tube as a control.
2. Place another tube in a beaker of hot water (over 90 °C). Record what happens.
3. Place another tube in a beaker of crushed ice and water. Record what happens.

Results:

(ii)

Observations on adding water	
Observations on adding concentrated hydrochloric acid	

(iii)

Observations on heating the boiling tube	
Observations on cooling the boiling tube	

Conclusions and comments:

_____

_____

_____

_____

(b) The equilibrium between $Cr_2O_7^{2-}$ and CrO_4^{2-}

Theory:

A solution containing chromate(VI) ions reacts as follows in the presence of acid:

$$Cr_2O_7^{2-}{}_{(aq)} + H_2O_{(l)} \rightleftharpoons 2CrO_4^{2-}{}_{(aq)} + 2H^+{}_{(aq)}$$

orange   yellow

This reaction is suitable for illustrating Le Chatelier's Principle because very definite colour changes are involved. Adding an acid will increase the concentration of H^+, and adding a base will reduce it.

This experiment is used to demonstrate the effects of **concentration changes** on an equilibrium mixture.

Chemicals and apparatus:

- Potassium chromate solution ☠
- Dilute sulfuric acid (2 M) ✖ i
- Potassium dichromate solution ☠
- Sodium hydroxide solution (2 M) 🜍
- White paper

- Boiling tubes and racks
- Dropping pipettes
- Beaker (100 cm³)
- Safety glasses

Procedure:

NB. Wear your safety glasses, and gloves.

1. Quarter fill a boiling tube with the solution of potassium dichromate provided. This should have an orange colour.

2. Keep a second sample of the potassium dichromate solution in a test tube as a control.

3. Carefully add some dilute sodium hydroxide solution until the colour changes. Record what happens.

4. Carefully add dilute sulfuric acid until the colour changes. Record what happens.

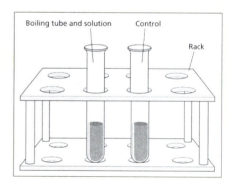

Results:

Observations on adding sodium hydroxide solution	
Observations on adding sulfuric acid	

Conclusions and comments:

_____

_____

_____

_____

(c) The equilibrium between Fe^{3+} and $Fe(CNS)^{2+}$

Theory:

A solution of iron(III) chloride reacts with a solution of thiocyanate ions as follows:

$$Fe^{3+}_{(aq)} + CNS^-_{(aq)} \rightleftharpoons Fe(CNS)^{2+}_{(aq)}$$

yellow   red

Adding hydrochloric acid reduces the concentration of Fe^{3+} by forming a complex ion containing iron and chlorine. This causes a shift of equilibrium to the left. The equilibrium can be shifted to the right hand side by adding some potassium thiocyanate solution.

Chemicals and apparatus:

- Concentrated hydrochloric acid
- Iron(III) chloride solution (0.05 M)
- Potassium thiocyanate solution (0.05 M)

- Boiling tubes and racks
- Dropping pipettes
- Safety glasses

Procedure:

NB. Wear your safety glasses.

1   Mix together about 5 cm³ respectively of solutions of iron(III) chloride and potassium thiocyanate in a boiling tube. Record what happens.

2   Divide the mixture into three portions in separate boiling tubes. Keep one of these as a control.

3   Using a fume cupboard, add some concentrated hydrochloric acid to the second tube. Record what happens.

4   Add an equivalent amount of water to the third tube. Record what happens.

5   To the second tube, add some potassium thiocyanate solution. Record what happens.

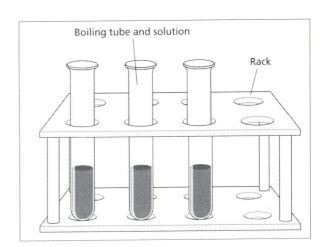

Boiling tube and solution

Rack

Results:

Solutions added to the boiling tubes	Observation
Iron(III) chloride and potassium thiocyanate	
Concentrated hydrochloric acid	
Water	
Potassium thiocyanate	

Conclusions and comments:

_____

_____

_____

_____

Title:_____**Date:**_____

Chemicals and apparatus:

_____

_____

_____

Procedure:

_____

_____

_____

_____

_____

_____

_____

_____

_____

_____

_____

_____

_____

_____

_____

_____

_____

_____

_____

_____

_____

Results:

(ii)

Observations on adding water	
Observations on adding concentrated hydrochloric acid	

(iii)

Observations on heating the boiling tube	
Observations on cooling the boiling tube	

Conclusions/comments:

_____

_____

_____

_____

Signature: _____

Student report, Activity 24a – diagram

Title:_____**Date:**_____

Chemicals and apparatus:

_____

_____

_____

Procedure:

_____

_____

_____

_____

_____

_____

_____

_____

_____

Results:

Observations on adding sodium hydroxide solution	
Observations on adding sulfuric acid	

Conclusions/comments:

_____

_____

_____

_____

Signature:_____

Title:_____**Date:**_____

Chemicals and apparatus:

_____

_____

_____

Procedure:

_____

_____

_____

_____

_____

_____

_____

_____

_____

_____

_____

_____

_____

_____

_____

_____

_____

Risk management:

_____

_____

_____

Results:

Solutions added to the boiling tubes	Observation
Iron(III) chloride and potassium thiocyanate	
Concentrated hydrochloric acid	
Water	
Potassium thiocyanate	

Conclusions/comments:

_____

_____

_____

_____

Signature: _____

Student report, Activity 24c – diagram

1   When sodium hydroxide solution is added to sodium dichromate solution, a colour change from orange to yellow occurs. Explain why this happens.

_____

_____

_____

2   What is observed when potassium thiocyanate solution is added to a solution containing iron(III) ions? Explain why this happens.

_____

_____

_____

3   When a solution containing cobalt(II) chloride, deionised water and hydrochloric acid is placed in a mixture of crushed ice and water, its colour changes from violet to pink. Explain why this happens.

_____

_____

_____

4   What is observed when a test tube half full of a violet solution containing cobalt(II) chloride, deionised water and hydrochloric acid is placed in a beaker of hot water? Explain why this happens.

_____

_____

_____

5   What is observed when concentrated hydrochloric acid is added to a test tube containing potassium thiocyanate solution and iron(III) ions? Explain why this happens.

_____

_____

6   What colour change would you expect to happen if dilute sulfuric acid is added to potassium chromate solution? Explain your answer.

_____

_____

_____

7   What is observed when potassium thiocyanate solution is added to a test tube containing potassium thiocyanate solution and iron(III) ions? Explain why this happens.

_____

_____

8   What is observed when hydrochloric acid is added to a test tube half full of a violet solution containing cobalt(II) chloride, deionised water and hydrochloric acid? Explain why this happens.

_____

_____

_____

9   What is observed when water is added to a test tube half full of a violet solution containing cobalt(II) chloride, deionised water and hydrochloric acid? Explain why this happens.

_____

_____

_____

10   Why is a control used in each of these experiments?

_____

_____

Activity 25 (mandatory experiment) – Determination of free chlorine in swimming pool water or bleach using a colorimeter or comparator

(a) Determination of chlorine using a colorimeter

Theory:

A colorimeter may be used to estimate the free chlorine in swimming-pool water or bleach. When chlorine compounds are added to swimming-pool water, the active agent is usually chloric (I) acid, HOCl. It kills micro-organisms by oxidation. Chloric (I) acid and its conjugate base, the chlorate(I) ion, ClO⁻, together make up what is called 'free chlorine'. These species react with iodide ions in solution in the same way as chlorine itself does, oxidising them to iodine:

$$Cl_{2(aq)} + 2KI(aq) \rightarrow I_{2(aq)} + 2KCl_{(aq)}$$

The more concentrated the chlorine in water, the more intense the colour of the iodine solution formed. The concentration of chlorine in a sample of swimming pool water or diluted bleach is obtained as follows. The colour obtained on reaction of the sample with potassium iodide solution is compared with those colours obtained by the reactions with potassium iodide solution of some solutions of known concentration of chlorine. A calibration curve is used in doing this.

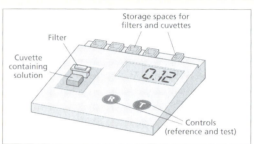

Chemicals and apparatus:

- 2 % potassium iodide solution
- 5 % ethanoic acid solution in water
- Sample of swimming pool water or diluted bleach
- Milton sterilising fluid
- Deionised water
- Colorimeter
- 440 nm wavelength filter

- Cuvettes
- 50 cm³ volumetric flasks with stoppers
- Burettes
- 250 cm³ volumetric flask
- 10 cm³ graduated cylinder
- Safety glasses

Procedure:
NB. Wear your safety glasses.

1   Dilute 2.5 cm³ of Milton Sterilising fluid to 250 cm³ with deionised water, using a 250 cm³ volumetric flask.

2   Add 5 cm³ of 5 % ethanoic acid solution to each of five 50 cm³ volumetric flasks, which are labelled A, B, C, D, and E respectively.

3   Use a burette to transfer the diluted Milton solution to the volumetric flasks as follows: 1.0 cm³

to flask B, 2.0 cm³ to flask C, 4.0 cm³ to flask D and 8.0 cm³ to flask E.

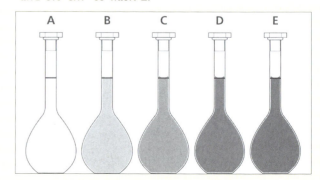

	Flask A	Flask B	Flask C	Flask D	Flask E
5 % ethanoic acid	5 cm³	5 cm³	5 cm³	5 cm³	5 cm³
2 % potassium iodide	5 cm³	5 cm³	5 cm³	5 cm³	5 cm³
Diluted Milton	0 cm³	1 cm³	2 cm³	4 cm³	8 cm³
Concentration of NaOCl in ppm	0	4	8	16	32
Total volume in flask	50 cm³	50 cm³	50 cm³	50 cm³	50 cm³

4   Transfer about 5.0 cm³ of 2 % potassium iodide solution i.e. an excess of KI, to each of the five flasks.

5   Dilute to the mark with deionised water in each case.

6   Stopper each flask and mix thoroughly.

7   Allow about five minutes for the free chlorine to oxidise the iodide ions to iodine. Colours of different intensity will result.

8   Switch on the colorimeter,

9   Place a 440 nm wavelength filter in the filter slot.

10 Pour each of the five standard solutions into a cuvette, rinsing each cuvette first with the solution it is to contain.

11 Zero the colorimeter in accordance with the manufacturer's instructions.

12 Measure and record the absorbance of each solution.

13 On the graph paper provided, plot a graph of absorbance versus concentration (in terms of chlorine) for the five standard solutions.

14 Add 5 cm³ of 5 % ethanoic acid solution, and then 5.0 cm³ of 2 % potassium iodide solution to a 50 cm³ volumetric flask, labelled F.

15 Fill the flask up to the mark with the sample being tested (swimming pool water or diluted bleach).

16 Allow about five minutes for the colour to develop.

17 Measure and record the absorbance for the solution in flask F.

Results:

Sample	Concentration of NaOCl in ppm	Absorbance
A	0	
B	4	
C	8	
D	16	
E	32	
F		

Concentration of free chlorine in the sample in flask F = _____

Conclusions/comments:

_____

_____

_____

_____

(b) Determination of chlorine using a comparator

Theory:

Free chlorine oxidises DPD No. 1 tablets (N,N-diethyl-p-phenylenediamine in the form of its sulfate) to a water-soluble red product. The more concentrated the chlorine in the water, the more intense is the red colour produced.

Chemicals and apparatus:

- DPD No.1 tablets ✖ n
- Sample of swimming-pool water or diluted bleach solution
- Comparator
- Safety glasses

Procedure:

NB. Wear your safety glasses.

1 Use some of the water being tested to rinse the compartments of the comparator.
2 Fill each compartment with fresh portions of the sample.
3 Add a DPD No.1 tablet to each compartment.
4 Crush each tablet with a stirring rod.
5 With the lid fitted, shake the comparator until the tablets have dissolved completely.
6 Compare the colour produced in the sample with the pre-calibrated standards built into the comparator.
7 Choose the best colour match.
8 Read the free chlorine concentration of the selected standard in p.p.m. (mg l^{-1}).

Results:

Concentration of free chlorine in the sample = _____

Conclusions/comments:

_____

_____

_____

_____

Title:_____**Date:**_____

Chemicals and apparatus:

_____

_____

_____

Procedure:

_____

_____

_____

_____

_____

_____

_____

_____

_____

_____

_____

_____

_____

_____

_____

_____

_____

_____

_____

_____

_____

_____

_____

_____

_____

_____

_____

_____

_____

Results:

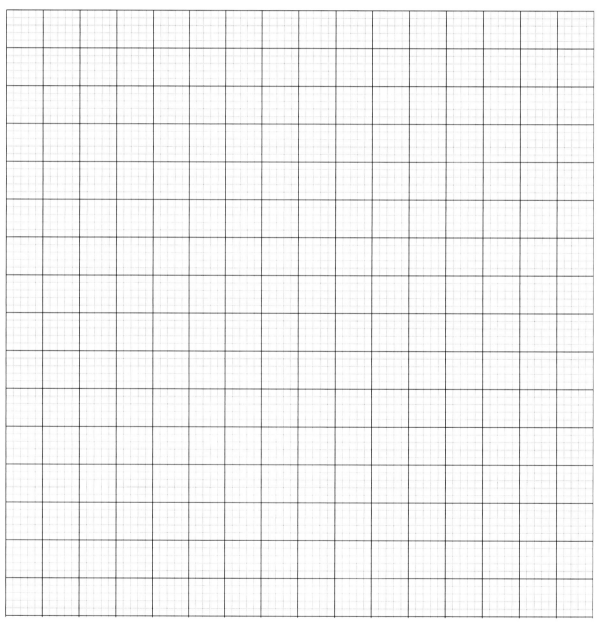

Conclusions/comments:

_____

_____

_____

_____

Signature:_____

Student report, Activity 25 – diagram

1   What is the connection between the colour intensity of a solution and the concentration of the dissolved coloured substance?

_____

_____

2   What does a colorimeter measure?

_____

_____

3   Name the two species that make up "free chlorine" in a sample of swimming pool water.

_____

_____

4   Why is potassium iodide used in the colorimeter experiment?

_____

_____

5   Why is it necessary to use excess potassium iodide in the colorimeter experiment?

_____

_____

6   Why is ethanoic acid used in the colorimeter experiment?

_____

_____

7   In using a comparator, why should it be rinsed with the water sample being tested?

_____

8   Why should the comparator compartments be filled exactly to the graduation marks?

_____

_____

9   In the comparator experiment, what precaution should be taken with the stirring rod before the crushing operation?

_____

10 In the comparator experiment, why should the DPD tablets be completely dissolved before comparison with the standards?

_____

_____

Activity 26a (mandatory experiment) – Determination of total suspended solids (expressed as ppm) in a water sample by filtration

Theory:

Amongst the problems encountered in natural water is turbidity, a cloudiness or lack of clarity caused by suspended particles. These are insoluble substances that are too finely divided to settle to the bottom, but are dispersed throughout the water sample. The cloudiness may reduce light penetration in surface water and interfere with photosynthesis in aquatic plants. These substances may also affect fish life or they may indicate sewage discharges. The total suspended solids in a water sample may be estimated in the laboratory by filtration, the mass of the filter paper being found before and after the sample is filtered.

Chemicals and apparatus:

- Water sample
- Small-pore filter paper
- Filter funnel
- Large beakers (1000, 500, or 250cm³)
- Conical flask
- Drying oven
- Electronic balance
- Safety glasses

Procedure:

NB. Wear your safety glasses.

1   Find the mass of a dry filter paper.
2   Filter 1 litre of the water sample.
3   Dry the filter paper in an oven.
4   Find the mass of the dried filter paper.

Results:

	Values (specify units)
Mass of dry filter paper	
Mass of dried filter paper after filtration	
Mass of suspended solids	
Volume of water sample	

Calculations:

Total suspended solids in mg/l (ppm)	

Conclusions/comments:

_____

_____

_____

_____

Title:_____**Date:**_____

Chemicals and apparatus:

_____

_____

_____

Procedure:

_____

_____

_____

_____

_____

_____

_____

_____

Results:

	Values (specify units)
Mass of dry filter paper	
Mass of dried filter paper after filtration	
Mass of suspended solids	
Volume of water sample	

Calculations:

Total suspended solids in mg/l (ppm)	

Conclusions/comments:

_____

_____

_____

_____

Signature:_____

Activity 26b (mandatory experiment) – Determination of total dissolved solids (expressed as ppm) in a water sample by evaporation

Theory:

Dissolved solids can affect the colour or taste of water. If the amount of dissolved solids is very high, it may be an indication that the sample is of salt water.

The total dissolved solids in a water sample may be estimated in the laboratory by evaporation, the mass of the beaker used in the experiment being found before use and after the water is evaporated. The increase in mass of the container, which equals the mass of dissolved solids, is then determined.

Chemicals and apparatus:

- Filtered water samples
- Graduated cylinder (100 cm³)
- Beaker (250 cm³)

- Drying oven
- Electronic balance
- Safety glasses

Procedure:

NB. Wear your safety glasses.

1   Find the mass of a clean dry 250 cm³ beaker.
2   Add 100 cm³ of a filtered water sample to the beaker.
3   Evaporate to dryness in an oven.
4   Find the mass of the dry beaker when cool.

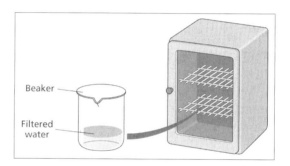

Results:

	Values (specify units)
Mass of beaker	
Mass of beaker and dissolved solids	
Mass of dissolved solids	
Volume of water sample	

Calculations:

Total dissolved solids in mg/l (ppm)	

Conclusions/comments:

_____

_____

_____

_____

Title:_____**Date:**_____

Chemicals and apparatus:

_____
_____
_____

Procedure:

_____
_____
_____
_____
_____
_____
_____
_____

Results:

	Values (specify units)
Mass of beaker	
Mass of beaker and dissolved solids	
Mass of dissolved solids	
Volume of water sample	

Calculations:

Total dissolved solids in mg/l (ppm)	

Conclusions/comments:

_____
_____
_____
_____

Signature:_____

Activity 26c (mandatory experiment) – Determination of pH of a water sample

Theory:

Extremes of pH can make water unpleasant to taste. It can also make the water corrosive, as well as affecting plant and animal life.

The pH of the water samples may be measured using a pH meter or pH paper or universal indicator solution. Data-logging pH sensors are also available, which can feed data directly to a computer or calculator.

Chemicals and apparatus:

- Water samples
- pH meter or pH paper or universal indicator solution
- Safety glasses

Procedure:

NB. Wear your safety glasses.

1   Calibrate the pH meter and dip the electrode or probe into the water sample. Record the steady reading.

2   Alternatively, add a few drops of universal indicator or dip a strip of pH paper into the water sample. Compare the colour obtained with the appropriate colour charts. Record the pH of the matching colours.

3   Repeat for each water sample.

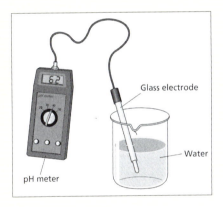

Results:

Water sample	pH

Conclusions/comments:

_____

_____

_____

_____

Title:_____**Date:**_____

Chemicals and apparatus:

_____

_____

_____

Procedure:

_____

_____

_____

_____

_____

_____

_____

_____

Results:

Water sample	pH

Conclusions/comments:

_____

_____

_____

_____

Signature:_____

1   What problems are associated with a high total suspended solids level in water samples?

_____

_____

2   What problems are associated with a high pH in a water sample?

_____

_____

3   What problems are associated with a high total dissolved solids level in water samples?

_____

_____

4   Why are filtered water samples used in the determination of total dissolved solids?

_____

_____

5   How are particles that cause a high total suspended solids reading removed in water treatment?

_____

_____

6   Suggest a possible cause of high levels of total suspended solids in water.

_____

_____

7   A volume of 500 cm³ of water was found to contain 0.07 g of dissolved solids. Express the concentration of the dissolved solids in ppm.

_____

8   Suggest a possible reason for high levels of total dissolved solids in water.

_____

_____

9   A volume of 500 cm³ of water was found to contain 0.025 g of suspended solids. Express the concentration of the suspended solids in ppm.

_____

_____

10   Suggest a reason for a relatively low pH in a water sample.

_____

_____

Activity 27 (mandatory experiment) – Estimation of the total hardness of a water sample using edta

Theory:

Hardness in water is caused by dissolved calcium and magnesium ions, $Ca^{2+}_{(aq)}$ and $Mg^{2+}_{(aq)}$. The total hardness of a water sample is found by titrating the sample with a standard solution of ethylenediaminetetraacetic acid (edta). The equation for the reaction between edta (H_2Y^{2-}) and calcium and magnesium ions (which are represented as M^{2+}) is:

$$H_2Y^{2-}_{(aq)} + M^{2+}_{(aq)} \rightarrow MY^{2-}_{(aq)} + 2H^+_{(aq)}$$

The indicator Eriochrome Black T is used to detect the end point. At the end point, the indicator changes colour from wine-red to dark blue.

The titration results are used to calculate the total hardness of the water sample.

Chemicals and apparatus:

- Hard water sample
- 0.01 M edta solution ✗ n
- Buffer solution (pH 10) 🗲
- Eriochrome black T indicator ✗ i
- Deionised (or distilled) water
- Burette (50 cm³)
- Retort stand
- Clamp
- Filter funnel
- Conical flask (250 cm³)

- Pipette (25 cm³)
- Pipette filler
- Wash bottle
- White card
- White tile
- Beakers (250 cm³)
- Graduated cylinder (10 cm³)
- Spatula for solid indicator
- Safety glasses

Procedure:

NB. Wear your safety glasses.

1 Rinse the burette with the edta solution, the pipette with the hard water, and the conical flask with deionised water.

2 Fill the burette to the mark with the edta solution, making sure that the part below the tap is filled.

3 Pipette 50 cm³ of the hard water sample into the conical flask, and add 2–3 cm³ of the buffer (pH 10) solution.

4 Using the spatula, add the solid indicator to the flask in minimal quantities, with swirling, until a deep wine-red colour is obtained.

5 Carry out one rough titration, and a number of accurate titrations until two titres agree to within 0.1 cm³.

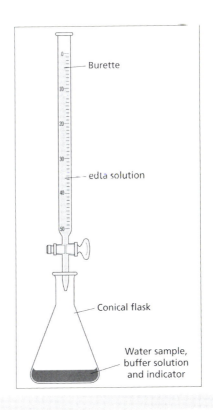
Burette
edta solution
Conical flask
Water sample, buffer solution and indicator

Results:

	Values (specify units)
Rough titre	
Second titre	
Third titre	
Average of accurate titres	

Calculations:

Volume of hard water sample	
Molarity of edta solution	
Average volume of edta solution	
Total hardness in mol/l Ca^{2+}	
Total hardness in g/l CaCO$_3$	
Total hardness in ppm CaCO$_3$	

Conclusions/comments:

_____

_____

_____

_____

Student report, Activity 27

Title:_____**Date:**_____

Chemicals and apparatus:

_____
_____
_____

Procedure:

_____
_____
_____
_____
_____
_____
_____
_____
_____
_____
_____
_____
_____
_____
_____
_____
_____
_____

Results:

	Values (specify units)
Rough titre	
Second titre	
Third titre	
Average of accurate titres	

Calculations:

Volume of hard water sample	
Molarity of edta solution	
Average volume of edta solution	
Total hardness in mol/l Ca^{2+}	
Total hardness in g/l $CaCO_3$	
Total hardness in ppm $CaCO_3$	

Conclusions/comments:

_____

_____

_____

_____

Signature: _____

Student report, Activity 27 – diagram

1 Why is it necessary to use a standard solution of edta in this experiment?

_____

_____

2 What is different about the procedure used in rinsing the conical flask prior to a titration in this experiment compared to that used when rinsing the burette or the pipette?

_____

_____

_____

3 What colour change takes place at the end point in each titration?

_____

4 Why is the funnel removed from the burette after adding the edta solution?

_____

_____

5 In using the burette in this experiment, why is it important (a) to clamp it vertically, (b) to make sure that the part below the tap is full of the edta solution?

_____

_____

6 What steps are taken to ensure that the pipette delivers the correct volume of the hard water sample into the conical flask?

_____

_____

7 What is the function of the buffer solution in this experiment?

_____

_____

8 Why is an indicator necessary in this experiment?

_____

_____

9 Describe the steps involved in determining the permanent hardness in the water sample you used in this experiment.

_____

_____

_____

Theory:

The dissolved oxygen present in a water sample can be measured by an iodine/thiosulfate titration, using the Winkler method. Under alkaline conditions manganese(II) sulfate produces a white precipitate of manganese(II) hydroxide:

$$Mn^{2+}_{(aq)} + 2OH^-_{(aq)} \rightarrow Mn(OH)_{2(s)}$$

This is then oxidised by the dissolved oxygen in the water, forming a brown precipitate:

$$2Mn(OH)_{2(s)} + O_{2(aq)} \rightarrow 2MnO(OH)_{2(s)}$$

The mixture is then acidified with concentrated sulfuric acid. Under these conditions, the Mn(IV) oxidises the iodide ions to free iodine:

$$MnO(OH)_{2(s)} + 4H^+_{(aq)} + 2I^-_{(aq)} \rightarrow Mn^{2+}_{(aq)} + I_{2(aq)} + 3H_2O_{(l)}$$

The free iodine is then titrated with standard sodium thiosulfate solution using starch solution in the usual way as the indicator:

$$I_{2(aq)} + 2S_2O_3^{2-}_{(aq)} \rightarrow 2I^-_{(aq)} + S_4O_6^{2-}_{(aq)}$$

Overall:

1 mole O_2 → 2 moles $MnO(OH)_2$ → $2I_2$ → 4 moles $S_2O_3^{2-}$

i.e. the ratio of dissolved oxygen to thiosulfate is 1:4.

Chemicals and apparatus:

- Manganese(II) sulfate solution ☒ n
- Alkaline potassium iodide solution 🧪
- Concentrated sulfuric acid 🧪
- 0.005 M sodium thiosulfate solution
- Starch indicator solution
- Deionised or distilled water
- Water sample
- Reagent bottle (250 cm³) with stopper
- Basin
- Burette (50 cm³)
- Retort stand
- Clamp
- Filter funnel
- Conical flask (250 cm³)
- Pipette (25 cm³)
- Pipette filler
- Wash bottle
- White card
- White tile
- Beakers (250 cm³)
- Droppers
- Graduated cylinder (10 cm³)
- Safety glasses

Procedure:

NB. Wear your safety glasses.

1 Rinse a 250 cm³ reagent bottle with deionised water, shaking vigorously to wet the inside and so avoid trapped air bubbles.

2 Immerse the bottle in a basin of the water sample so as to fill the bottle, making sure that there are no trapped air bubbles.

3 Add about 1 cm³ each of manganese(II) sulfate solution and of alkaline potassium iodide solution to the bottle, using a dropper placed well below the surface of the water.

4 Stopper the bottle so that no air is trapped – some water will overflow at this point.

5   Invert the bottle repeatedly for about a minute, and then allow the brown precipitate to settle out.

6   Carefully add 1 cm³ of concentrated sulfuric acid to the bottle, by running the acid down the side of the bottle.

7   Restopper the bottle, being careful not to trap any air.

8   Redissolve the precipitate by inverting repeatedly. If all the precipitate has not dissolved at this point add a little more acid and repeat the mixing process. The brown colour of iodine should now be visible.

9   Rinse the pipette, burette and conical flask with deionised water.

10 Rinse the pipette with the iodine solution and the burette with the sodium thiosulfate solution.

11 Pipette 50 cm³ of the iodine solution into a conical flask and titrate with 0.005 molar sodium thiosulfate solution.

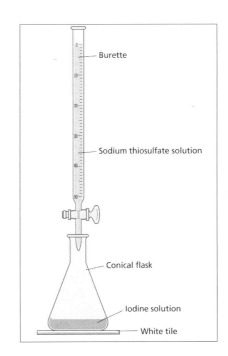

12 Add about 1 cm³ of starch indicator when a pale yellow colour is present. A dark blue colour is produced and the titration is continued until this colour just disappears.

13 Carry out a number of accurate titrations until two titres agree to within 0.1 cm³.

Results:

	Values (specify units)
Rough titre	
Second titre	
Third titre	
Average of accurate titres	

Calculations:

Volume of water sample	
Molarity of thiosulfate solution	
Average volume of thiosulfate solution	
Concentration of dissolved oxygen in moles/l	
Concentration of dissolved oxygen in g/l	
Concentration of dissolved oxygen in ppm	

Conclusions/comments:

_____

_____

_____

_____

Title:_____**Date:**_____

Chemicals and apparatus:

_____
_____
_____

Procedure:

_____
_____
_____
_____
_____
_____
_____
_____
_____
_____
_____
_____
_____
_____
_____
_____
_____
_____
_____
_____
_____
_____
_____
_____
_____
_____
_____
_____
_____

Results:

	Values (specify units)
Rough titre	
Second titre	
Third titre	
Average of accurate titres	

Calculations:

Volume of water sample	
Molarity of thiosulfate solution	
Average volume of thiosulfate solution	
Concentration of dissolved oxygen in moles/l	
Concentration of dissolved oxygen in g/l	
Concentration of dissolved oxygen in ppm	

Conclusions/comments:

_____

_____

_____

_____

Signature: _____

Student report, Activity 28 – diagram

1 Why is a standard solution of sodium thiosulfate needed in this experiment?

2 Why are the reagent bottles completely filled with the water sample?

3 What colour change is observed when manganese(II) sulfate solution and alkaline potassium iodide solution are added to the water sample in the conical flask?

4 Why are two accurate titrations carried out in this experiment?

5 At what stage of each titration is the starch indicator added to the conical flask?

6 What colour change takes place at the end point in each titration?

7 State one safety precaution that you would take in this experiment.

8 During each titration, the walls of the conical flask are occasionally washed down with deionised water. Why is this done?

9 How closely should the accurate titration results agree with each other?

10 Describe the steps involved in measuring the biochemical oxygen demand of a water sample.